Wildlife Survivors
The flora and fauna of tomorrow

John R. Quinn

TAB Books
Division of McGraw-Hill, Inc.
Blue Ridge Summit, PA 17294-0850

1 2 3 4 5 6 7 8 9 0 FGR/FGR 9 9 8 7 6 5 4 4

Library of Congress Cataloging-in-Publication Data
Quinn, John R.
      Wildlife survivors : the flora and fauna of tomorrow / by John R.
Quinn.
         p.   cm.
      Includes bibliographical references (p. 202) and index.
      ISBN 0-8306-4346-X (H)   ISBN 0-8306-4345-1 (p)
      1. Adaptation (Biology)—North America.   2. Natural selection-
-North America.   3. Man—Influence on nature—North America.
I. Title.
      QH546.Q56 1993
      574.5'097—dc20   93-21037
                                                          CIP

Acquisitions editor: Kimberly Tabor
Editorial team: Joanne Slike, Executive Editor
                Susan Wahlman Kagey, Managing Editor
                Susan Bonthron, Editor
Production team: Katherine G. Brown, Director
                Wanda S. Ditch, Layout
                Joan Wieland, Proofreading
                Joann Woy, Indexer
                Ruth Gunnett, Computer Artist
Design team: Jaclyn J. Boone, Designer
                Brian Allison, Associate Designer
Cover design: Denny Bond, East Petersburg, Pa.   4353
Cover illustration: John R. Quinn   TEK2

Dedication

For my mother—a feisty, adaptable member of the genus *Homo*, species *sapiens*, and a survivor, if there ever was one.

Acknowledgments

Any work penned by an amateur field naturalist and wildlife artist that concerns itself with the great biological and ecological transformations currently occurring on our continent must rely heavily on the work of other, more specialized observers of the natural scene. The popular press, in particular *The New York Times*, *The Wall Street Journal*, *The Asbury Park Press*, and the (Bergen County, NJ) *Record* among daily newspapers, and *Natural History*, *Fisheries*, *Bioscience*, and *Animals' Agenda* among nature and science-oriented periodicals, have been important sources of current environmental events and trends. With respect to individuals, I have borrowed freely and gratefully from the work of the scientists, naturalists, nature writers, and lovers of the outdoors whose authoritative and comprehensive books are listed in the bibliography and to whom I owe a considerable debt. In addition, numerous friends and colleagues, of both lay and scientific association, have offered advice, counsel, pertinent published material, and personal observations on our changing flora and fauna. In particular I thank biologist Jerry G. Walls for contributed materials, and my brother Steve, of the American Museum of Natural History, for West Coast faunal observations. To Kim Tabor and the gang at TAB Books must go thanks for their patient fielding of my frequent and often voluminous correspondence on the subject; and to my computer-wise wife, Luci, for patiently guiding me through the complex and ever-changing environment of a new personal computer, and in the process making a survivor out of one particular author.

About the author

John R. Quinn has been an artist-naturalist for more than 40 years, specializing in the aquatic environment and its myriad creatures, from fishes to marine mammals. He has had numerous one-man shows of his wildlife art and has written nine books on nature, science, and outdoor subjects. In the past he has been associated with the exhibits department of the Academy of Natural Sciences of Philadelphia, and has served as exhibits designer for The Science Center of New Hampshire and science curator and public relations director for The Bergen (NJ) Museum of Art and Science. He also operated the Quinn-Life Studio, a New Hampshire-based exhibits design and nature art concern. He is currently staff artist for TFH Publications, a major publisher of pet and animal books.

Quinn is an active diver and snorkeler and has collected fishes and other marine life for aquaria and scientific study. He divides his time between Avon by the Sea, New Jersey, on the famous Jersey Shore, and rural Craig County, Virginia.

Contents

Prologue

A changed world

There isn't any real doubt that major changes are inevitable. The only question is what kind. Some of them will happen no matter what we do, because as the environment deteriorates they will happen automatically. Others will be brought about by our efforts to stave off disaster. All the changes will be significant, and the world of the next generation is going to be quite different from our own.

Frederik Pohl, *Our Angry Earth* (Tom Doherty Associates, New York, 1991)

Biologists have long believed that humankind, in its ancient beginnings a woodland/savannah creature, has not done itself any favors by vastly restructuring its world and surrounding itself with the cluttered, crowded, noisy, suburban and urban environment as we know it today. Population biologist Paul Ehrlich likens life under the present environmental circumstances to forcing urban humanity to "fly with one wing," against the grain, so to speak, of our collective genetic inheritance.

The situation is not likely to improve in the near future. Sometime near the end of this decade, mankind will pass a milestone of sorts: for the first time in recorded history, more people—about 53 percent—will live in and around cities than in rural areas. The populations of major urban areas worldwide, including the United States, will, according to educated estimates, undergo a dramatic rise that will result in an age of 21 world "megacities," each with populations of 10 million or more. Mexico City already has double that figure and Calcutta is home to some 12 million inhabitants. Some of Africa's major cities are growing at the astounding rate of 10 percent per year.

With the earth's human population growing at the rate of about 100 million per year, the forces driving the wave of urbanization are almost irresistible. Los Angeles, with a population of 11.8 million people in 1992, is expected to swell to 13.2 million by the end of this decade; by contrast, New York has 16.2 million today and is expected to reach "only" 16.6 by the year 2000. In the first decade of the 21st century, the majority of the human beings living in North America will live out their lives under conditions that cannot be called anything other than urban.

As long as the three great biospheres of the earth—the land, the water, and the atmosphere—continue to cycle the oxygen, nitrogen, and carbon from the clouds to the land and the seas and back again, there will always be living things in relative and considerable abundance upon the planet. Life perseveres, even in the face of implacably hostile environmental conditions. Virtually all living organisms possess some degree of adaptability and resilience—the ability to react to, and to effect, the few essential changes in survival-oriented behavior in response to external stimuli. In spite of all the profound changes mankind has wrought upon the earth, its foundational elements yet retain the ability to nurture and sustain life. None of us would be drawing breath at this moment if this were not the case. We, in our burgeoning numbers, function reasonably adequately within the framework of the declining world biotope today, as do a vast host of fellow travelers on spaceship earth. But the world community becomes smaller and increasingly monotypic in more ways than one. What kinds of habitats will we see more of in the future? The question might well be asked: "Where do you live, and have you noticed physical changes or increased population pressures in your own area's environment?"

Scott Russell Sanders writes in his wonderful book, *Townships*:

> Home ground is the place where, since before you had words for such knowledge, you have known the smells, the seasons, the birds and beasts, the human voices, the houses, the ways of working, the lay of the land and the quality of the light. It is the landscape you learn before you retreat inside the illusion of your skin. You may love the place if you flourished there, or hate the place if you suffered there. But love it or hate it, you cannot shake free. Even if you move to the antipodes, even if you become intimate with new landscapes, you still bear the impression of that first ground.

Be that as it may, the near future portends an urban "home ground" awaiting most of us, including the myriad other organisms that share this planet—and in particular, this continent—with us. To date, "slash and burn" urbanization—population growth and development of a region until it no longer offers life of any discernible quality, followed by emigration to other areas—continues apace. Rural and natural lands are disappearing at a frightening rate near all of the continent's major urban centers, due primarily to unrestricted immigration and development. As long as this unregulated growth and all-consuming destruction of the carrying capacity of the living landscape continues, the impoverished biospheres of the r-selected, survivor-type organisms can only expand until they include the length and breadth of the North American continent—"from sea to shining sea."

Introduction

And I brought you into a plentiful country, to eat of the fruit thereof and the goodness thereof; but when ye entered, ye defiled my land and made mine heritage an abomination.

Jeremiah 2:7

The environment of a post-extinction landscape favors what biologists call r-selected creatures—that is, species that are highly mobile, adaptable, and opportunistic. In our world these are represented by rats, roaches, sparrows, gulls, and weeds.

Christopher Manes, *Green Rage* (Little, Brown; 1990)

It doesn't require the vision of a seer or the perception of a Rhodes Scholar to realize that the living landscape that surrounds and nurtures us is, and has been for some time, undergoing vast, profound change. Change is a perfectly natural consequence of life; it occurs constantly and at all levels of physical existence, even in the most pristine environments, anywhere on earth.

Great tides of change have flooded the earth many times in its five-billion-year history. Entire biospheres have come and gone, and with them, many myriad life forms. Most of the biological transformations of the prehistoric past occurred over vast periods of time, often over hundreds of millions of years. Today, physical change is once again making itself felt, but unlike those natural and glacially slow metamorphoses of the past, the winds moving over the planet as the third millennium nears are fast, powerful, and unpredictable. Whether they are blowing fair or foul remains to be seen, but the winds of change in the waning years of the 20th century are altering the world's habitats profoundly.

To gain some sort of idea as to the world that plants and animals—especially humankind—will occupy in the years and the decades that lie ahead, take a good look at the United States from the air, preferably at night. Not long ago, I flew from the New York metropolitan area to Portland, Oregon, via a stop in Atlanta, Georgia. The trip turned out to be a great deal more than a simple transcontinental hop to visit family members on the West Coast; it was a graphic and unsettling education in rampant futurism of the first magnitude, on levels both majestic and mundane.

The view of much of this continent, at night and from the air, is at once one of awesome splendor and terrifying portent. From horizon to horizon, the spangled, glistening threads of the megalopolis are cast ever farther abroad into the darkness of the countryside, so that even where the aircraft passes high above regions distant from the sprawling conurbations of the coasts, multitudes of winking lights and the macroscopic headlights of lilliputian cars pierce the darkness below.

Gazing from the pressurized comfort of an aircraft cabin at this vast, astounding light show that now dominates much of the earth's surface, I cannot help but marvel at the energy, industry, and sheer genius that went into the rapid conversion of a once primeval planet into what is today truly a "global village." Millions of miles of roads connect an ever-spreading physical infrastructure, high-intensity lighting and all, into giant, brightly lit "world townships" visible from the moon.

One is also forced to wonder just what effect this immense transformation of the natural into the machine has on the multitudes of other beings that call the planet home—those whose required environments do not consist of asphalt plains or the domesticated and managed landscapes of the spreading cities. How do the birds and beasts fare among the lights? Among the many thousands of square miles of parking lots, freeways, shopping malls, townhouse communities, theme parks, golf courses, discount department stores, oil refineries, and neighborhood fast food outlets?

Scanning the night horizon, atwinkle from end to end with the brilliance of civilization, from a 727 flying at 32,000 feet, I found the question both intriguing and troubling.

But the true genesis of the book you are about to read actually came about through a seemingly insignificant event that took place on *terra firma* and much closer to home—the poignant sight of a lone, white cabbage butterfly flitting erratically across a lush green lawn on a typical suburban street in New Jersey on a hot summer day.

The sighting occurred in the course of a pleasant and most relaxing walk I took in the company of my wife one wonderfully bright and beautiful July day a couple of years ago. We strolled just for the pleasure of it, going nowhere in particular, just exploring on foot the immediate environs of our middle-class, suburban neighborhood. This pedestrian mode of travel was in itself rather unusual, for we, like almost everyone else in suburban society today, traditionally rely on the automobile to get from one place to another in the course of daily life.

But here we were, actually walking about and taking more direct notice of a neighborhood we knew well, though for the most part through the windshield of a car. I had never felt a need or reason to stroll these tree-lined streets. As an artist/naturalist, I saved my physical exertions for the wild places I loved, such as isolated beaches and canoeable rivers often distant from the suburbia most of us call home—or at least isolated from it, as in parks and protected nature reserves.

On this particular day, I was suddenly prompted by the sight of the little white butterfly lilting airily over the grass to take a closer look around me and note the inventory and nature of "things natural" that crossed our path. I took stock of the environmental makeup of my neighborhood through the eyes of an ecologist (for the first time, I'm ashamed to say), and I was at once surprised and rather disturbed by what I discovered.

Before going further, let me tell you a little about my neighborhood, if for no other reason than to assure you that its characteristics are shared by many, if not most, population centers throughout North America, and that I am not some privileged nature writer barricaded in his log cabin and bemoaning the gradual invasion of his splendid, forested isolation.

We live in a moderately large, well-kept, 20-year-old garden apartment complex in New Jersey's Monmouth County. Rentals here are not bargain basement, nor is the entire region a cultural or economic backwater by any stretch of the imagination. Monmouth, roughly 60 miles from midtown Manhattan and one of the famous Jersey shore counties, experienced rapid and explosive population growth throughout the booming 1980s, so that what was, a mere decade or two ago, an essentially rural, bucolic landscape terminating in the east at the state's famous coast is rapidly becoming an urbanized infrastructure of upscale housing developments, shopping malls, corporate business parks, and the maze of cluttered, ever-busy traffic arteries that link and serve them. In this respect it is typical of many localities throughout the United States and southern Canada. Virtually no area of the continent has escaped the effects of more than 250 million citizens seeking a home in the country and the good life that goes with it.

Monmouth County today is pretty much a new suburban complex, and the resident flora and fauna have been profoundly affected by the habitat fragmentation and loss that accompanies this level of rapid and intense development. My own apartment complex is flanked on the north and west sides by narrow swaths of woodland that appear reasonably intact; to the east by a major divided highway; and to the south by light industrial and commercial enterprises. Directly across the busy highway lie the 100-odd acres of a yet-undeveloped county park, its peaceful, densely wooded glades and unpaved pathways offering respite from the surrounding racket of suburbia.

This suburban park's forested sections abound with bird life during the spring and fall migrations, but at other times of the year they are strangely empty and silent. Its uncut fields harbor a riot of spring and summer wildflowers but seemingly few butterflies and other larger, more conspicuous insects. Although the woodlands look green and lushly normal, something's missing, something I took for granted not all that many years in the past. Visible and audible animate life seems sparse, very much like the environs of the surrounding suburb.

My ground-up inspection of my own neighborhood resulted in a distressingly meager tally of visible resident organisms. Beginning with the grass cover itself, the lawns are broad, lush and green, sun-dappled in tree shadow and well watered by ample summer rainfall, but they consist primarily of domesticated rye grass and fescue, and a few hardy "weed" invaders like plantain, crabgrass, white clover, and dandelion—an impoverished floral biotope if there ever was one.

The makeup of the local shade tree population seems, at first look, to be as diverse as any forest. A closer look reveals that roughly a dozen species predominate in the suburban environment today, as the less attractive "scrub" species, such as many of the oaks, are removed and exotics are introduced.

Mobile organisms are in abundance in my neighborhood, but a second look reveals that they are a plethora of individuals of rather few species. House sparrows and starlings are everywhere, superseded in number only by street pigeons and that aggressive little newcomer from the American West, the house finch. To be sure, mockingbirds, mourning doves, robins, and the occasional crow can be seen on any street in town, and catbirds and cardinals patrol the streetside shrubbery, but the riot of bird sights and sounds that were a taken-for-granted feature of the suburban landscape a mere 20 years ago seem to have faded into silence in the thin, faintly hazy air.

With the exception of the ubiquitous feral cats, mammals of any kind are few and far between, at least during the daylight hours. Raccoons, opossums, and skunks still survive in fair numbers in my part of New Jersey, for every new day brings a fresh crop of roadkills that surely must put a dent in the ranks of the survivors. As the enclaves of natural land here have become more and more fragmented and isolated from one another, hundreds of small mammals and birds meet death on woodland trailways that are now bisected by rivers of asphalt and swiftly moving machines. The roads of today are physical barriers as impenetrable and effective as a raging stream; virtually uncrossable in most urban situations, they divide and conquer, isolate and ultimately destroy the living landscape they intrude upon.

And so, year by year, the richly varied tapestry of nature unravels—a little here, a little there. Even here, close to home, the variety of living creatures—and the many little encounters with them that we once took for granted—subtly but inexorably decrease. Their ranks are replaced by hardier, somewhat less poetic wildlife better equipped to deal with the new world we've created. Paper wasps, yellow jackets, earwigs, and houseflies replace the butterfly legions that once populated these woods and fields; the ethereal twilight fugues of the wood thrush and whippoorwill fade within the shrinking cathedrals of the vanishing woodlots, replaced by the tuneless chirpings and screechings of the European sparrow and the starling.

Little by little, as I saw in the course of a single stroll down a normal street on a normal summer day, the world had changed. The truly frightening thing about this sudden perception is how unthinkable it was a mere 25 or 30 years ago. A late afternoon stroll through one's neighborhood then would have yielded many more interactions with organisms of multiple and varied stripes, from pesty mosquitoes to warbling orioles and nations of grasshoppers and butterflies in every weedy lot. No one gave a thought to the outlandish possibility that some day, in the not-too-distant future, all of these myriad creatures that we once took for granted would fade slowly from the realm of our collective experience, under a sun and sky that had suddenly become dangerous enemies themselves.

The forces that have caused this grim, inexorable transformation, which we seem unable to stem or influence in any meaningful way, are many. Some are sudden and violent, others insidious and very patient. Unless you know what to look for or are sensitive to this sort of thing, you might miss the signs entirely. You might gaze at the vibrant greenness all about you, feel the timeless touch of the wind and the power of the sun, listen to the voices of birds and children, and wonder what all the fuss is about. Life goes on, so they say, but what of the fate of its visual and sensual and biological richness? How will the quality of the human experience fare in an altered and artificial landscape populated by those living creatures—human and nonhuman—that can adapt to it and survive? This might be our appointed fate as things look now, but is it inevitable?

The threats to the ultimate survival of plant and animal species worldwide today are many and varied and, surely by now, well-known to everyone. Some of the advancing perils are direct and violent, others subtle but nonetheless of great import. They include:

- *habitat destruction and degradation*, including urbanization, deforestation, desertification, and the resulting soil erosion.
- *overhunting and overfishing*, in particular of wetland-dependent waterfowl and of declining foodfish stocks and marine mammals.
- *herbicides, pesticides, and other toxic compounds* entering the food chain.
- *global warming* and its disruption of world weather patterns and entire ecosystems.
- *acid rain* and other forms of air pollution.
- *toxic waste buildup* and environmental contamination.
- *ozone depletion.*
- *surface water and groundwater pollution* and their great threat to entire aquatic ecosystems.
- *overcollecting of animals* for pets and other wildlife products.
- *poaching* and other uncontrolled illegal activities.
- *introduction of exotic species.*
- *hazardous human refuse* such as plastics and other solid waste.
- *burgeoning human populations*—the root cause of all environmental decline.
- *disturbance and stress* caused by contact with humans while involved in birding, hiking, boating, and other nonconsumptive forms of recreation.
- the gradually *deteriorating quality of the ecosphere* that all of these factors combine to produce.

In Aristotle's time, in the fourth century B.C., some 500 species of organisms were recognized by the science of the period. Today, somewhere in the neighborhood of 1.5 million species of plants, animals, fungi, and microorganisms have been described and classified by science and assigned Latin names. Each year, scientists the world over describe several thousand plant species and more than 10,000 animal species, the latter mostly small invertebrates. Of the known species today, some 265,000 species of plants and about 45,000 species of vertebrate animals have been studied to one degree or another and are thus relatively well known to educated people.

Two decades ago it was speculated that at least five million species of organisms—the majority of them insects and microorganisms yet undiscovered—call the planet earth home. Studies conducted in the world's rain forests throughout the 1980s have concluded that the total is probably much higher, perhaps somewhere between 30 and 80 million! The ongoing and catastrophic destruction of the bulk of the tropical ecosystems, where more than half the world's biomass is stored, portends the destruction, by the year 2025, of at least two million species out of a conservative, middle-ground estimate of 10 million species extant today—a terrifying prospect.

The Geneva-based International Union for the Conservation of Nature and Natural Resources has further estimated that more than 30,000 of the known species of plants and animals are in a precarious state worldwide. Their listing includes not only those classed as threatened and endangered, but also species simply considered rare and vulnerable as a result of human activities or direct campaigns of persecution. By the most conservative estimates, nearly 2000 species worldwide merit real concern as to their chances of survival if present trends continue. In the United States alone, 492 plant and animal species are currently listed as either threatened or endangered, and 3600 species from around the world are under consideration for federal protection and import prohibitions. Roughly 400 species are presently on the Fish and Wildlife Service's top-priority "Candidate I" list for immediate protection under the terms of the Endangered Species Act. Whatever "doomsday list" you prefer to take seriously, the outlook for the vast majority of the world's living things—all too few of them displaying the stern stuff of survivorship—is not bright.

In sum, we are headed almost inevitably for the third of the three great extinction spasms of recent earth history. The first occurred about 65 million years ago and resulted in the abrupt disappearance of the dinosaurs and other reptiles. The second happened some 12,000 years ago and witnessed the swift demise of most of the large land fauna of North America. It has been speculated that early hunters were responsible for the complete destruction of the great herds of North American herbivores during the last Ice Age, and if this is so, the present looming environmental crisis is but the next step in the process. Scientists have calculated that during the "great dying" of the dinosaurs 65 million years ago, these animals vanished at the rate of about one species every 10,000 years. This is a far cry from the one species per day (or even per hour, depending on the researcher) reportedly slipping into extinction today.

But all is not lost.

Those resilient plants and animals that will be left to us following the great vanishings of the coming decades will be the survivors, the so-called r-selected, or "weedy species" of the present age.

The word survive itself stems from the ancient Latin word supervivere—super meaning "over," and vivere meaning "to live." Thus, a survivor is one who continues to live or exist, usually over great odds or after some event or destructive force that has caused the end of the existence of all of the others. We normally think of survivors as being people who escaped death in some human-caused or natural disaster, such as a major conflict in war, an airplane crash, a fire, or an earthquake. When a person dies of an illness, those family members left behind, who go on living, are often referred to as "the survivors." In other words, for most of us the word survivor is usually applied to people, to a person who manages to live through a terrible ordeal or who regains health following a severe illness.

However, possessed of life as they most assuredly are, animals and plants can be called survivors as well. Even a microbe can overcome tremendous adversity and emerge triumphant. On the individual level, an animal that manages to escape the clutches of a predator can certainly be said to have survived the attack. A deer that is injured in a fight with another deer or struck by a hunter's bullet and then escapes and later heals can also be said to have survived a life-threatening situation. Many living things can also survive indirect threats, including such adverse environmental conditions as extreme heat or cold, prolonged drought or flooding, or other drastic changes in their living space, or habitat. Those plants or animals that can go on living after facing such threats to their existence are certainly survivors.

The plant and animal survivors I'll discuss in this book are not those that have managed to escape being killed and eaten by some other creature at some point in their individual lives. All organisms exist for a certain time and must die eventually. Rather than individual survivorship, I'll look at the survival of entire genera and species of plants and animals in the face of the great and often destructive changes taking place all over the earth today, and most especially, here in North America.

It is a biological fact that many of the earth's plants and animals cannot survive what scientists call modifications to the habitat. Some of these environmental modifications might be the result of natural events, such as the great damage inflicted by a powerful hurricane on coastal areas, or the destruction of an entire forest ecosystem by a lightening-caused fire. Such natural disasters might cause the disappearance or death of many individual plants and animals living within the affected area, but they usually do not result in the extinction of an entire species unless that particular species is endemic to the habitat, or lives only in that area.

But things have changed drastically since the dawn of the Industrial Revolution some 200 years ago. Since then, profound changes have been taking place over the earth at an ever-accelerating rate due to the spreading and intensifying activities of humankind. Some of these changes have not affected plants and animals very much, while others have had a heavy impact upon many species of virtually all plant and animal phyla.

Most of the modifications humans have made to natural habitats virtually everywhere in the world have been carried out to provide living space and food for humans or to extract the resources necessary for producing artifacts of the culture. When people were not as numerous as they are today, the earth and its creatures could adjust to the alterations people made in the environment. At the time of Christ, for example, only about eight million people lived on earth, most of them concentrated in cultural centers in the Near and Middle East. Population concentrations then were few and far between. People were widely scattered over the earth 2000 years ago, and their lifestyle and technology was very primitive compared to that of today. The cumulative effect of their activities on the environment was not one of intense and widespread degradation of natural systems. The healing capabilities of the planet could more easily absorb and rectify scattered and sporadic ancient assaults made upon the global ecology when humankind numbered less than one billion souls.

By 1970, however, 3.7 billion people called the earth home and occupied all of the land area of the planet with the exception of Antarctica and the vast, forbidding deserts of North Africa. Today our numbers have risen to nearly 5.4 billion, and by the year 2000, population scientists believe roughly 6.3 billion people will live on earth, nearly double the number that were alive only 23 years ago! And still the end is nowhere in sight: the earth's human population is expected to double again within the next half-century.

Even this huge, almost unthinkable number of people might not have an utterly catastrophic effect upon the earth's vast but fragile life support system if most of these folk were subsistence farmers employing primitive agricultural methods, as was the case 2000 years ago. But our industrial technology, even in the developing Third World, involves hundreds of chemicals and poisons that are used in everything from the manufacture of countless industrial products to the fertilization of crops and the killing of insect pests. Moreover, in the industrialized world we now have machines that can clear land, stop and alter the flow of rivers, and construct buildings and highways on a far greater scale than mere muscle power ever could. But the problem is one that sorely afflicts not only the affluent West but the entire global village, for it is estimated that today at least 40% of the earth's total land surface is intensively utilized by humankind and another 25% to 30% is less intensively exploited.

The result of all of this frenetic activity, carried out every hour of every day, all over the earth, by those 5.4 billion people struggling to feed, clothe, and house themselves and their families, has been massive and often destructive changes in the global environment. Most of these alterations to the environment mean that a forest, marsh, prairie, or even a desert is modified by technology and transformed into a domesticated landscape more suited to human use. A forest must be leveled before homes, factories, or highways can be built. A prairie containing dozens of plant species is overgrazed by domestic cattle or is plowed under and planted in a monoculture of corn, wheat, or one of about 10 other major food plants. The desert environment, in its natural state at least, simply cannot support the burden of numerous people and their complex infrastructure of homes, roadways, shopping centers, and lush green lawns, so a river is dammed and its water diverted to the dry land so that people can live there in a lifestyle completely at odds with the natural cycles of the entire region.

In many respects, the story of the worldwide decline in biodiversity and the resultant species extinction is a numbers game revolving around the increasing number of our own species and the activities we pursue. Figures pertaining to habitat destruction, in particular those concerning worldwide deforestation, are among those numbers most familiar to environment-conscious people everywhere. In contemplating the growing specter of deforestation, however, a vague, false sense of complacency creeps in via the perception of the distance of the threat. It is common knowledge that tropical rain forests are being leveled—even as you read these words—at a rate of roughly a square mile every 10 minutes, or 70 acres per minute. This amounts to an area of tropical forest the size of the state of Illinois, or about 55,000 square miles, being stripped and plundered of its natural vegetation and resident animal biomass every year. At this frightful rate, scientists speculate that all but a few isolated blocks of intact moist tropical forest will have been destroyed within the next 25 years, with the remainder disappearing by the middle of the next century.

Rain forest destruction is all the more disturbing when you consider the fact that the level of diversity to be found in this habitat is unequaled anywhere else on earth. A 30-acre parcel of Indonesian rain forest was found to harbor more than 700 species of trees, about the same number as are found on the entire North American continent. A recent survey by the New York Botanical Garden found 450 tree species, 13 of them completely unknown to science, on a 2½-acre parcel of Brazil's Atlantic coast forest. A mere 10 species occupy the same area in a North American forest. A single tree in a Peruvian forest was found to be home to 46 species of ants, as many as are found throughout the British Isles!

The destruction of tropical ecosystems, horrifying enough in itself, is made all the more so by the knowledge that they contain an estimated half of the world's known species of plants alone. If merely half of these threatened species disappear when the forests have been reduced to less than 10% of their original extent—possibly by the end of the century—then perhaps 65,000 plant species, or one-quarter of the world's total will have been destroyed forever. Indeed, scientists predict that more than half of the world's entire species roster will be doomed to extinction by the middle of the next century if current trends are not halted or reversed.

But though these grim enumerations of environmental destruction and extinction evoke deep feelings of horror and helplessness in North Americans who care at all about the state and fate of their world, there is some small degree of comfort to be had in the remoteness of the carnage. After all, "Brazil is thousands of miles away, and that lovely stretch of woodland that I love so dearly is still there and things don't seem to be all that bad in my state or province." You can still get away from it all to the green quiet and the healing waters of nature when you need to, right?

Well, perhaps—but if so, for how long?

Consider another set of figures a little closer to home. In 1992, a bad year overall for the American home construction industry, close to a million acres of the natural landscape was irrevocably altered to suit the needs of our growing population, now tallied at about 255 million people. Most of us experience feelings of resigned dismay when we witness the destruction of one small, perhaps personally familiar natural Eden, sacrificed in the course of the construction of new homes or yet another strip mall. The sight of the destruction, so close to home and thus intensely personal, seems somehow mitigated by the belief, no matter how wistfully delusionary, that this is only an isolated and perhaps unique loss of natural terrain of no great magnitude and that "there are still plenty of woodlands elsewhere." But again, the figures say otherwise. The key is to be found in the term, *housing starts*, used by the home construction industry to indicate new construction of single-family dwellings and apartment houses in a given period. In November 1992, the home building industry expressed concern because housing starts in the Northeast were down 0.8 percent, to an annual rate of 124,000 units. The annual rate is determined by multiplying monthly sales data by 12, taking into consideration seasonal variations. Nationwide, the industry had recorded a 3.2 percent decline for the month of June 1992, with 1.17 million units begun, down from 1.21 for the previous month. The West recorded a 303,000 annual rate; the Midwest, 277,000; and the South 463,000, all showing declines from higher, more optimistic figures.

By late December, however, the picture had begun to change, reflecting, industry analysts said, "a gradual recovery from the persistent recession." In that month 1,242,000 single-unit housing starts were reported nationwide, with construction begun on 109,000 structures with five or more living units each—in other words, condos and townhouses.

The point of all this developers' arithmetic, which does not take into account the construction of shopping malls, industrial parks, airports, and the like, is found in a simple and direct question: For how long can a continent even as roomy as North America continue to accommodate such a volume of construction activity without sustaining permanent and irreversible environmental damage? In the past, no one gave much thought to the question of what happened to the plants and animals that lived in natural areas when humans and their machinery arrived and changed the habitats forever. Most people either gave the matter no thought, or believed that the plants were replaceable with cultivated vegetation and that the wild creatures would simply "go somewhere else." For a long time, that attitude was valid because there was, indeed, plenty of room in North America, and clean air and water and plants and wildlife seemed to be unlimited in abundance. But today we know we are rapidly running out of "somewhere elses," and with ever more people in the world who must be fed and clothed and housed, a multitude of plant and animal species are being pushed into the twilight zone of endangerment.

Although the survival outlook for many of the world's plant and animal species is far from rosy, no matter how much the land is changed to suit the needs of people there will always be some plants and animals that can adapt to the new environment and live there. This book is not about endangered species, but rather about those organisms that will likely replace the former as they vanish from the earth, many of them in our own lifetimes. We mourn this terrible loss, progressing and accelerating on nearly every biological front every moment of every day, but there is little we can do but mourn without the stabilization of human populations worldwide.

This book also faces a few incontrovertible facts: In spite of the burgeoning memberships of conservation and environmental organizations worldwide, despite all the eloquent pleas for restraint and mercy in our dealings with the earth and its creatures, despite the pronouncements of politicians and the good intentions of the people they lead, the battle against global environmental degradation is being lost. More and more people are gravely concerned and actively committed to doing something about the looming crisis, but faced with a grim and relentless litany of cross-accusations and denials, a blizzard of environment versus growth catch-phrases and the horrifying electronic and printed imagery of ecocide in near and far places, they wonder whether the individual commitment to recycle cans and bottles and turn off the water while brushing is enough. Will these noble but seemingly symbolic and insignificant acts really help stem the large-scale destruction of tropical rain forests or the ozone layer, or heal the dying seas, or return the world's life-giving atmosphere and weather patterns to preindustrial normality?

This book does not attempt to answer those questions. Their answers reside in the future alone, and I can only speculate on the prognosis based on the current trends hypothesis. I do, however, try to make an educated guess—based on observations we all can make and information readily available to all—as to the floral and faunal makeup of this continent in the coming decades.

Some of the questions I ask and attempt to answer involve both the whys and the hows of animal survivorship. That is, what are the forces, factors, and conditions that caused the environments of the particular survivors discussed here to change, and why were these organisms—these r-selected creatures—able to adapt to those changes and thrive while other plants and animals could not. The answers to these questions will surprise, disturb, entertain, and, perhaps most of all, help give you a better understanding of how our earth's environments function when they are under stress, and why, long after the last spotted owls, sturgeons, whooping cranes, whales, and wood warblers have become a wistful memory, you'll still see plenty of raccoons, tent caterpillars, robins, and starlings around your suburban neighborhood and catch plenty of carp, eels, and tough little shiners in the less-than-pure rivers. They have simply made the best of a profoundly and perhaps permanently changed world; they are coming to terms with the biologically impoverished landscapes of tomorrow, and will survive and thrive in spite of them— indeed, perhaps because of them.

These long-term environmental troopers, then, perhaps much more so than those creatures that endure and persevere over short-term odds, are the true wildlife survivors of the coming century.

Note: The survivors discussed in this book by no means constitute the entire roster of organisms that will likely persevere and even thrive in the altered North American environment of tomorrow. There is every possibility that a few of the species discussed might decline in the face of future changes —both natural and human-caused—in the environment that might ultimately threaten their survival. Likewise, organisms currently occupying only a minor foothold on the continent might burgeon in numbers in response to suddenly favorable circumstances in their environments. The sudden and ominous appearance and spread of the exotic Asian tiger mosquito and the zebra mussel are apt cases in point.

My selection of survivors was made on the basis of the organisms' current levels of abundance, familiarity to the majority of citizens through either personal contact or the media, and obvious signs that these hardy and adaptable flora and fauna show every indication of strengthening and expanding their respective environmental niches in the future. In short, nearly everybody knows that carp, ragweed, starlings, raccoons, and gypsy moths seem to get along fine within the complex infrastructure of 20th century mankind, and show no inclination to be shunted aside or eradicated like their more sensitive, specialized, and less fortunate brethren. The question is, "Why?"

Wildlife Survivors attempts to answer that question.

Chapter one

The new environments

We are rapidly filling the world with ourselves. As a result, the rest of the living world is forced aside to make room for more of us. The earth is rapidly losing valuable plant and animal life. If trends continue at this rate, by the middle of the next century the number of extinct species could exceed those lost in the great extinctions of the geologic past.

John Erickson, *Dying Planet* (TAB/McGraw-Hill, 1991)

"North America of today is one of the most fully developed continents of the earth. No comparable area is as productive. None has a higher standard of living...it has the main sinews of modern industry." These glowing and optimistic lines were penned by a respected Canadian geographer in 1963, some eight years before the first Earth Day and the advent of the so-called "Green Revolution." They reflect the prevailing temper of the times, an innocent age when economic and industrial growth, unencumbered by regulation or suggestions of restraint, was regarded as sacrosanct and considered far above even the mildest condemnation and reproach—long before the word *development* took on the ominous overtones it has acquired today.

Grade school geography texts published into the early 1970s featured photographs of smoke-belching industrial complexes and heavy machinery carving more and more miles of superhighways into the scenery side by side with pristine wilderness vistas full of picturesque wildlife, as though these two widely divergent aspects of 20th-century America were somehow perfectly compatible with each other. The inference was that there would always be room on this great continent to accommodate an infinite expansion of the former within the finite expanses of the latter, no matter how crowded with people the land became.

In the post-colonial year of 1800 people seriously imagined that it would take between 500 and 2000 years to settle and fully develop North America.

We have, of course, learned otherwise. Within about 100 years the economic and population growth rates of the United States have exploded to about 10 times that of Canada and five times that of Mexico.

Historically speaking, the public awakening to the impending environmental crises of the next century has occurred within the blink of an eye. A mere generation ago the notions that sunbathing could kill you, or that commonplace creatures like frogs, migratory songbirds, and minnows were vanishing across the length and breadth of the continent, or that the world's forests were dying under the combined assault of chain saws and industrial acids were completely absent from the public consciousness. People then regarded the world around them pretty much as they always had, as a varied, sometimes beautiful, sometimes ugly place that would remain as it had been for ages—composed of the often ugly and stress-filled conurbation of industrial civilization, countered by a wealth of distant but accessible postcard-pretty natural places in which to escape the end results of technology.

Today, although the environmental doomsayers are often disbelieved by a sizeable percentage of the population, the collective human consciousness has become indelibly stamped with the vaguely unsettling knowledge that all is not right with the earth. We are somehow aware that although the sun rises each day in an apparently normal sky and—so far—no monstrous and insoluble social or environmental calamity has descended upon us in force, great and profound changes are in the wind for all of us, and they are not likely to be pleasant ones.

Although many of the environmental calamities and setbacks of recent years seem distant and removed from us here in North America, the reality of former Interior Secretary Stewart Udall's "quiet crisis" has come home to roost for those of us aware of the great changes that have taken place in our immediate surroundings, no matter where we live. The horrific environmental destruction caused by the 1991 Gulf War, the huge oil spills occurring worldwide, the on-going destruction of the tropical rain forests, the slaughter of Africa's wildlife, all affect us deeply and profoundly, but not immediately—not physically and personally.

But most of us have watched with quiet dismay the gradual destruction and "uglification" of our "home ground," our own particular corner of the continent. Whether it be the insidious surrender of natural lands to highways and housing and shopping malls, the abrupt destruction of a familiar and beloved patch or woodland or meadow, or simply the increasing stress of attempting to coexist with too many of our fellows in a complex and confusing environment, we are forced to come to grips with the same environmental realities that now confront the "lesser beings" that share the planet with us.

The new environments, rapidly evolving in every part of the globe in the final years of the 20th century, all share two primary characteristics that qualify them as wildlife survivors' habitats: they have been moderately or severely modified by the presence or activities of humans; and they are moderately or densely populated by people gathered in either agricultural or urban communities.

The urban environment

The deterioration of the environment, both physically and esthetically, is most apparent in our cities.

Paul R. Ehrlich, *Healing the Planet* **(Addison-Wesley, 1991)**

The dictionary defines the word *urban* as "pertaining to, located in, or constituting a city." The word *city* can also be taken to mean a center of population, and here some interesting and eye-opening statistics on the rapid growth of the urbanized United States can be had. Shortly after the end of the Revolutionary War, in 1790, the population center of the young nation was placed at just east of Baltimore; by 1820 it had shifted west of the mountains in what is today West Virginia. By 1880 the area just west of Cincinnati, Ohio, had the honor, and today the population center lies in the area of Duluth, Minnesota.

The great urban centers of the United States today include the Boston-to-Richmond megalopolis, centered in New York; the Milwaukee-to-Michigan City complex, centered in Chicago; the Santa Barbara-San Diego conurbation, centered in Los Angeles; and the Dallas-Fort Worth area in Texas. The country as a whole has become spheres of influence revolving around the conurbations of New York, Chicago, and Los Angeles. There is little doubt that although there are yet large areas of moderately populated rural and generally undeveloped land in North America, the great majority of Americans today are urbanites. This was not always the case. In 1800, 93.9 percent of the population was rural. Today the figure has shrunk to about 22 percent, this great demographic change occurring as

Americans discovered over the past 200 years that the true "land of opportunity" was not "down on the farm," but among the lights of the burgeoning cities.

Contrary to popular belief, the urban environment does not consist solely of physical surroundings best exemplified by New York City's Times Square or Los Angeles' Wilshire Boulevard. Although unrelieved pavement and soaring, impersonal skyscrapers are indeed an integral and inseparable part of the inner urban scene, the bulk of the surrounding cityscape can comprise many square miles of lower-density residential areas, manufacturing and heavy industrial facilities, miles of major thoroughfares flanked by retail stores, and acre upon acre of empty lots, abandoned building sites, and waste areas.

The truly urban habitat is harsh and unforgiving when it comes to survival—whether that of the homeless human being or any of the urbanized plants and animals that exploit the urban environment's few "eco-niches." All of the environmental factors that supply the requirements for life in the healthy, natural ecosystem are absent, or very nearly so, at the city's core.

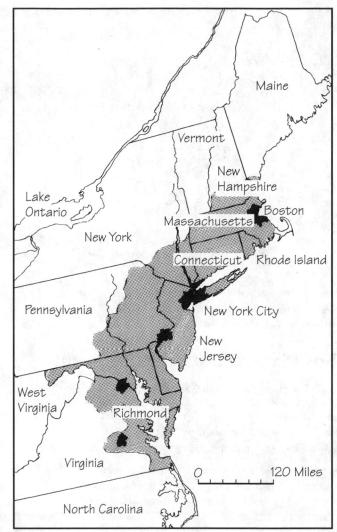

With the exception of urban parks, there is little undisturbed or vegetated terrain in the urban environment and thus many, if not most, of the organisms found there are heavily dependent upon people and their structures for food and shelter. Gray squirrels, starlings, and rock doves often abound in urban parks, but their numbers would quickly decline if affectionate and committed humans ceased feeding them. The contained and overused city parks are mere "habitats-in-a-bottle," and cannot, on their own, support the high bird and mammal densities that often exist there.

Only the hardiest of plant and wildlife survivors and feral creatures are able to persist in the heavily urbanized situation. A representative list of such survivors includes:

- **Plants:** Ragweed, goldenrod, milkweed, pokeweed, dandelion, staghorn sumac, plane tree, ailanthus, and crabgrass.
- **Insects:** Cockroach, ant, fruit fly, earwig, termite, silverfish, mosquito, and house fly.
- **Reptiles:** Brown snake.
- **Birds:** House sparrow, starling, and rock dove.
- **Mammals:** House mouse, brown and black rats, feral dogs and cats.

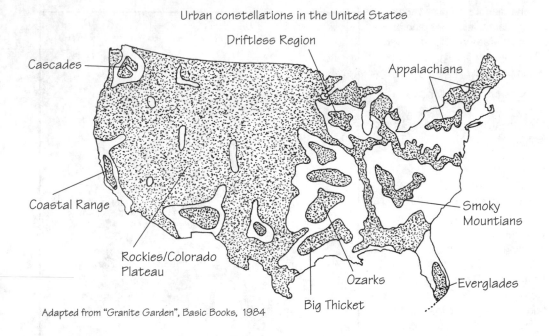

Urban constellations in the United States

Cascades

Driftless Region

Appalachians

Coastal Range

Rockies/Colorado Plateau

Smoky Mountians

Big Thicket

Ozarks

Everglades

Adapted from "Granite Garden", Basic Books, 1984

The older suburban environment

4

The man under forty now living in the American small town is living a life that would have seemed as strange as some fictional life on the planet Mars to his father when he was a young man. All the physical aspects of life in the towns have kept changing and changing. Life has taken on a constantly accelerated pace.

Sherwood Anderson, *Hometown* (Alliance Books, 1940)

The word *suburb* stems from the Latin, *suburbium*: *sub*, meaning "near," and *urbs*, meaning "city." As commonly envisioned today, the suburbs immediately surrounding a given metropolis consist primarily of established residential districts interspersed with the commercial and transportation infrastructures that service them. The older suburban environments ringing most metropolitan centers today are now a part of the city proper rather than the fringes they once were. As the vast housing developments, road networks, and commercial complexes of the new suburbs have sprawled further from the cities' cores, the older suburbs have been left behind, either to age gracefully or to fall into the sad decay summed up in the term "inner city."

In most cases, the older suburban environment offers adaptable plants and wildlife a viable habitat. The tree-lined residential streets and yards lush with ornamental greenery present not only an appealing visual image but an ecologically viable physical environment as well. Such environments are attractive to resident songbirds, and larger mammals such as raccoons, skunks, and opossums. The representative survivors of the old suburban environment include:

- **Plants:** Grasses, clovers, plantain, weeping willow.
- **Insects:** Field cricket, ant lion, cabbage and sulphur butterflies, red admiral butterfly, tiger swallowtail, monarch butterfly, paper wasp, hornets and yellow jackets, honeybee, praying mantis.
- **Reptiles and amphibians:** American and Fowler's toads, green anole.

- **Birds:** House finch, blue and Steller's jays, catbird, towhee, tufted titmouse, cardinal, mourning dove, ground dove, mockingbird, song sparrow, common grackle, American robin
- **Mammals:** Gray squirrel, cottontail rabbit, raccoon, opossum.

The new suburban environment

Are we really qualified to make important decisions about the land and its resources if we cannot even identify the birds, plants and trees of our yards that have been around us since birth?

Naturalist Jon Young

The new suburban environment is largely the result of the rapid "slash and burn" urbanization that has overtaken the outlying areas of most of the larger North American urban centers in the past 20 years. The phenomenon is commonly described by the term *urban sprawl*, and represents the negative and environmentally destructive aspects of rapid population growth and the attendant development of the physical infrastructure.

The enormous growth of the new suburban environment has had a profound and lasting effect upon plant and animal species intolerant of habitat destruction or disturbance due to vastly increased human use of the land. The new suburbs are generally composed of large-scale tract housing, extensive corporate parks and light-industrial complexes, vast shopping malls, and the improvement or extension of the existing transportation network. In recent years, particularly in the Northeast, Florida, and the far West, slash and burn urbanization has occurred so rapidly that the ecology of entire regions has been altered or destroyed within a decade or two.

In the ten or twenty years immediately following the complete urbanization of a region, nearly all native flora and fauna are severely reduced in number or wiped out. Plants and animals dependent on extensive and unbroken forest cover are among the first to disappear. They are followed by migratory species that decline as they fail to find adequate breeding territories, or as human and domestic animal disturbance increases over time. A woodland or desert scrub surrounded by development might appear intact, but it will be subject to constant invasion by pest species such as cowbirds, dogs, cats, raccoons and, yes, people—either on foot or in mechanical conveyances.

Representative survivors of the new suburban environment include:

- **Plants:** Queen Anne's lace.
- **Insects:** Japanese beetle, broad-winged katydid, ladybird beetle, bumblebee, cabbage butterfly.
- **Amphibians and reptiles:** Spring peeper, garter snake.
- **Birds:** Red-winged blackbird, robin, bluejay, house sparrow, ring-billed gull, turkey vulture, common and fish crows, Canada goose, mallard, ring-necked pheasant, red-tailed hawk.
- **Mammals:** Red-backed vole, muskrat, striped and spotted skunks, white-tailed deer, white-footed mouse, woodchuck.

Landfills and waste areas

Rats? You wanna see rats? I'll show you rats as big as house cats. Come out here at night with a light and you'll see all the rats you want.

Landfill compacter operator, Kearny (NJ), 1982

The sanitary landfill concept was adopted gradually on a small, localized basis. It is still used on a town-by-town basis throughout North America in rural areas. By the 1930s, however, large-scale, regional landfills were to be found on both coasts, primarily in the area of New York City and in southern California. The term *sanitary landfill* was reportedly coined by one Jean Vincenz, then Commissioner of Public Works in Fresno, California. By 1945, at least 100 American cities had adopted the landfill idea, and within 15 years, more than 14,000 landfills were in operation around the nation. Today, municipalities in the United States still generate 180 million tons of solid waste each year; the landfill as a physical and environmental entity is not likely to disappear in the near future. The Environmental Protection Agency predicts that landfills will still be receiving some 50 percent of the total waste volume in the year 2000, down from 73 percent in 1988.

The basic idea behind the sanitary landfill is simply to dump a day's worth of garbage into a hole in the ground, compact it with heavy machinery, and then cover the mess with a layer of fill dirt. Abandoned mining pits and other such cavities have been pressed into service as landfill sites, and in some areas trash is simply piled on the ground and covered, the resulting mound rising until it forms a sort of man-made garbage mountain.

Landfills served a dual purpose in the past: to get rid of garbage and at the same time to reclaim what was considered "worthless" wetlands and marshes. The notion of converting undeveloped marginal lands into productive real estate, while simultaneously disposing of urban trash, was the prime motivating force behind the landfill concept from the very beginning.

Most of the earliest landfills were placed on the worst terrain possible for the purpose—wetlands. The leaching of liquids and toxic materials from older, carelessly sited landfills into the surrounding water system has been the major drawback and the biggest complaint against the sanitary landfill as a garbage disposal agent. Today, the value of wetlands is more completely understood and appreciated, and nearly all modern landfill operations are sited over nonporous substrates that will more effectively contain the liquid and toxic materials dumped there.

For the most part, active landfill sites, that is, those with broad areas still under daily operation surrounded by many acres of soil cover and a dense, established weed crop, are primarily attractive to scavenging animals, invasive weed plants, and disease organisms. The list of representative survivors includes:

- **Plants:** Mullein, poison ivy, phragmites, ragweed, Japanese knotweed, kudzu.
- **Birds:** Gulls, starlings, red-winged blackbirds.
- **Mammals:** Rats, mice.

The fragmented forests

With the exception of the large tracts of forested lands under public stewardship, which are protected from outright destruction but "managed" for timber, mineral, and oil extraction and recreational use, much of the woodlands of North America can be considered "at risk" habitats in the future. Although some genera and species of native plants and animals benefited initially from the early conversion of forests into farms and grasslands, and from the subsequent return of abandoned farms into forest in this century, current population growth dictates that more and more undeveloped land is appropriated for housing and road building. Formerly thought of as a problem endemic to the populous East, natural lands are being bisected and isolated in nearly all sections of the United States today. The dry, upland pine forests of interior Florida and the desert forests of the arid Southwest are both being decimated at an alarming rate, and as the great urban complex extending from Boston to Richmond expands to the west, the eastern hardwood forests are increasingly nibbled away by residential and commercial development.

The fragmentation of forests by spreading urbanization and by the web of transportation networks has resulted in the decline of those animal species requiring large expanses of intact habitat. Many deep-forest birds, such as wood warblers and thrushes, have suffered in the face of growing disturbance by human hikers and casual explorers and by increased victimization by the parasitic cowbirds. Some of the larger mammals, in particular raccoons, foxes, and deer, are accustomed to wandering over a considerable area in search of food, mates, or new territories. As the road network expands virtually all over the lower North American continent, the effect of these manmade barriers to free wildlife movement has become one of increasing concern (see chapter 8, "The Roadkill Factor").

Fragmentation of habitats, be they forests, marshlands, or prairies, also contributes to the decline and

Forested areas remaining in North America

extinction of its occupant species (particularly small populations of those species) through less direct—though no less ominous—means. For example, the gene pool is reduced when a population is isolated from others of its kind, and although individual numbers of an organism might seem high, vulnerability to destructive inbreeding increases as some genes disappear while less desirable ones are passed on.

Once the numbers of an animal population fall below a certain number of individuals in an isolated habitat, the chances increase that most or all of the individuals in a succeeding generation will be either male or female. In addition, localized natural disasters, such as fire, floods, or violent storms can destroy the remnant of an already endangered species entirely.

Representative survivors of the fragmented forests include:

- **Insects:** Tent caterpillar, gypsy moth, underwing moth.
- **Amphibians:** Red-backed salamander.
- **Mammals:** Coyote.

Tomorrow's aquatic environments

Few if any major [U.S.] river systems are unaffected by the threat to ecological integrity.
   The New York Times, Jan. 23, 1993

The current grim state of the freshwater aquatic ecosystems of the North American continent are of great concern to biologists and environmentalists. Under the combined assault of urbanization, siltation, agricultural runoff, and acid precipitation, many of North America's rivers and streams, as well as entire drainages, have become seriously degraded and no longer support the diversity of fish life they once held. Nearly 400 North American fish species are considered imperiled today. Some of them are rather obscure nongame species with small ranges; others, such as the sturgeons and salmonids, are (or were) important food and game species of much wider distribution.

Since most of the continent's aquatic ecosystems are not isolated entities, but are interconnected over often vast areas, pollution and urban runoff can affect areas far from the source of the effluent. Thus the soil runoff from a clear-cut logging operation can affect salmon spawning beds hundreds of miles downstream, and the sewage effluent of a town or city can degrade a stream or river for its entire length.

Anyone can take note of the overall health of their region's rivers and streams simply by noting the appearance of watercourses as they pass beneath roads. In areas undergoing intense development, few streams run clear at any time of the year; they appear brown and murky with construction runoff often originating far upstream. It must be remembered that in a healthy ecosystem soil is rarely washed from the land into the aquatic drainage, even at times of spring floods. Only in places where the vegetation cover has been removed from the land will precious soil be lost to erosion and end up in waterways.

The long-term prognosis for North America's freshwater ecosystems is not good. Much of the damage sustained at this point in time is irreversible, and it is highly unlikely that most stream and river systems can ever be returned to their preindustrial state. Under the continuing pressure of population growth, it is unrealistic to think that commercial and residential development can be halted or even slowed appreciably; to do so would be counter to the desires of the people and every politician in office today. Thus it is almost certain that further degradation of rivers and streams will occur over much of the continent.

All of this portends a low-diversity continental aquatic ecosystem populated primarily by r-selected organisms—the aquatic survivors. They will consist of fishes and other hardy organisms tolerant of warmer waters carrying a heavy load of suspended debris and toxins, which do not require silt-free substrates and clean water for both survival and reproduction.

The list of representative aquatic survivors includes:

- **Plants:** Hydrilla, water hyacinth.
- **Fishes:** American eel, bowfin, common carp, goldfish, common shiner, golden shiner, mosquito fish, common sucker, rudd, sea lamprey, cunner, mummichog, oyster toadfish, walking catfish, bullhead catfish, pumpkinseed, bluegill, largemouth bass, brown trout, zebra mussel.
- **Amphibians and reptiles:** Green frogs and bullfrogs, snapping turtle, painted and map turtles.

Protected areas

We're losing resources . . . the sad news is: we're even losing them where we're trying to protect them.

National Parks Service spokeswoman Molly Ross (*Roanoke* (VA) *Times-World News*, 1992)

The continent's wildlife refuges, national and provincial parks and forests, and nationally and locally protected natural areas offer both large and small oases of relatively intact and undisturbed habitats surrounded by a growing urban and agricultural complex. The United States' National Wildlife Refuge system encompasses some 14 million acres, excluding Alaska. In the largest state, more than 77 million acres are included in the refuge system, a fourth of which is the 19.5-million-acre Arctic National Wildlife Refuge. Although this impressive acreage total is not likely to be reduced in area, or removed from federal protection and management and allocated to other uses in the near future, it is almost certain that pressures to develop the natural resources contained within public lands will intensify as privately-owned domestic supplies dwindle or are exhausted.

Commercial development, along with increased human "multiple recreational use" of wildlife refuges and natural areas can only have a deleterious effect on their biospheres. Today roughly 100 federal wildlife refuges are used for military training of one kind or another, from low-level aircraft strafing to tank maneuvers. Sport hunting is permitted on 259 of the 472 refuges, and 91 permit trapping. The estimated annual wildlife refuge hunting toll is about 400,000 animals, including waterfowl.

The major, long-term prognosis for most natural areas would have to be that as more sensitive organisms progressively decline and disappear under increased human disturbance and modification of the habitat, the r-selected organisms common to the urban and suburban ecosystems will move in and colonize these vast and unexploited new territories.

The developed coastline

Enjoy your own island paradise right here in New Jersey! An exciting new lineup of elegant waterfront homes . . . from only $289,900.

Real estate ad, 1993

The coastal areas of North America will almost certainly continue to urbanize if present population demographic trends continue. Projections indicate that by the year 2010 more than 50 percent of the population will live within 50 miles of the coasts. The notion of a home with a view of water has always had popular appeal. North American real estate located anywhere near salt or fresh water is among the highest-priced, beyond the reach of most people. This powerful financial incentive has served to make the implementation of effective regulation or outright prohibition of coastal development an uphill battle. The resulting continuous trend toward virtually unregulated coastal development has continued despite the obvious threat of storm and tide damage.

Many older, more established shore communities, particularly in the urban Northeast, occupy marginal and filled lands mere yards from the high-tide mark. The loss of coastal wetlands continues in spite of public awareness of the importance of these fragile and crucial habitats. Eighty per cent of the Hudson River's original estuarine wetlands have been lost to filling and development, according to the American Littoral Society; they add that the loss continues despite protective legislation and an apparent public desire to preserve them.

The long-term projections for these fragile and unique coastal environments do not bode well for those plants and animals unsuited for adaptation to the profound and drastic changes that will come with urbanization and increased human use of the littoral zones. Versatile and aggressive birds and mammals, such as gulls, fish crows, muskrats, and raccoons, will have little trouble securing breeding and feeding space on the urbanized coast, while those more specialized species that require intact and more isolated habitats—such as the critically endangered piping plover, as well as the least tern, diamond-back terrapin, and the various species of sea turtles—will almost certainly disappear as viable wild species.

Filled and reclaimed
marginal lands in the New York metropolitan area

Source: Adapted from *Waste Management*
Regional Plan Assoc. NY 1968

5 Miles

In addition to the continued threat of overdevelopment, the littoral and inshore marine environment also absorbs huge amounts of sewage effluent and solid wastes. A National Academy of Sciences study estimated that oceangoing vessels annually discharge some six million tons of solid wastes into the oceans. Ships might jettison up to 450,000 plastic containers and other items daily worldwide, adding to the land-based waste stream that ends up in the coastal environment. The eventual fate of the littoral environments will likely be that of scattered enclaves of rigidly protected, but heavily utilized natural recreational areas surrounded by the burgeoning suburbs of the great megacities and urban constellations of the future. Their primary floral and faunal residents will be many of the hardy and adaptable organisms discussed in this book. The representative coastal survivors include:

- **Plants:** Phragmites, goldenrod.
- **Birds:** Herring gull, ring-billed gull, laughing gull, mute swan, cattle egret.
- **Mammals:** Muskrat.

Farmland, airports, golf courses, industrial parks

The broad, open areas of farmlands and other human alterations of the natural landscape that have not involved the construction of buildings and the extensive paving of land surfaces have traditionally offered wildlife at least marginal habitats and food and shelter in the form of abundant weed crops. The less intense farming methods of the past have benefited many species of game birds and songbirds, and have provided good-to-excellent habitats for many mammal species. Airports and golf courses were once touted as prime breeding areas for such shorebirds as upland sandpiper and killdeer, and the broad lawns of corporate industrial parks still offer perfect grazing meadows for urbanized woodchucks, Canada geese, and hordes of blackbirds.

Whether these specialized, man-made environments continue to serve as wildlife habitats in the future remains to be seen. Farming, for example, has become a much more extensive, mechanized and "cleaner" enterprise, resulting in the repeated and intensified use of fields through fertilization, the elimination of hedgerows, and the heavy use of a wide array of pesticides in the control of insect pests.

Much of the farmland that still exists at the fringes of the spreading East and West Coast megacities will doubtless be appropriated for development as urban sprawl overtakes it. Organisms requiring larger territories and feeding ranges, such as the white-tail deer or the red and gray foxes, will likely be displaced or reduced in number as they increasingly come into contact with humans and their vehicles, or become the object of control campaigns as a result of the decline of regulated sport hunting.

The list of representative survivors includes:

- **Birds:** Ring-necked pheasant, turkey and black vultures.
- **Mammals:** red fox, cottontail, woodchuck, white-tail deer.

Chapter two

The plants

At the time of Christopher Columbus' "discovery" of the New World, the landmass that is today the United States of America was luxuriantly and wonderfully cloaked with about 800 million acres of virgin forest. Think of it—the entire nation was covered coast-to-coast by the same awesome, green cathedrals of the now-remnant old-growth forests we are trying desperately to save today.

Despite intensive forestry and reforestation efforts, by 1930 the American forest had shrunk to 100 million acres, and today the accepted figure is around 50 million acres. This does not imply that 50 million acres are covered with trees like the ancient, moss-covered monoliths hanging on by a spotted owl's feather in the Pacific Northwest, but rather lands having unbroken forest of any kind, even twice or thrice harvested over the past 200 years.

On the world scene, the plant picture is equally grim. At the end of the last century it was estimated that the planet's dry land area was composed of 42 percent forest, 34 percent desert, and 27 percent agricultural and grasslands. Today those figures are 33 percent, 40 percent, and 29 percent respectively, indicating that while the natural forest cover has declined, creeping desertification and the spread of agricultural and grazing lands has increased. In addition, it must be remembered that the term forest as used here doesn't necessarily mean stands of great trees the size and age of the "General Sherman" redwood in California, but rather those of marginal and cutover woodlands, as well as scrubland, chapparal, and stunted boreal and taiga forests. Forests, in one form or another, are estimated to cover about 10,000 million acres of the earth's surface today.

The plants as a group constitute about 85 percent of the earth's total biomass, with trees making up 90 percent of the vegetation biomass. While trees usually face the direct threat of destruction for lumber, firewood, or land clearing for agriculture or development, the shrubs, grasses, and herbs are more often eliminated by broader, though no less destructive forces. These include widespread use of herbicides, poor agricultural practices, increased human use of the habitat, and climatic trends due to global warming and tree removal.

The vulnerability and decline of the native North American flora can be difficult to determine and predict. Individual plants, stationary organisms that they are, can neither flee a persecutor nor abandon a habitat that has been altered or destroyed—they are obliterated right along with it. In 1975 the Smithsonian Institution submitted a list to Congress in which 3187 kinds of plants, more than 1000 of them from Hawaii, were considered likely candidates for endangered status. At present there are relatively few—perhaps 200—plant genera or species out of the 14,880 known species of higher plants that could be considered to be in imminent danger of extinction. Today, some 3000

North American plant species are considered rare and subject to further decline or extinction if their habitats are damaged or destroyed.

While the above figure might not seem monumental, especially when once considers that the orchids alone number 26,000 species, the plants are sensitive bellwethers of the environment. Where the "green beings" are plowed under and their habitats destroyed, so too are the habitats of the higher organisms that depend on them, on and on up the food chain—and ultimately, to us.

A number of plant species, such as the endangered curly grass fern and the state of Maine's famous Furbish's lousewort are of extremely limited or fragmented distribution. Some species have been reduced to single, vulnerable colonies—and in their case, the destruction of a major portion of the extant habitat could spell the end for them in the wild state.

The principal threats native plants face are some of those that imperil all living things on earth today—in descending order of importance: habitat destruction, acid precipitation, overharvesting of timber, atmospheric pollution, and the present suspected global warming trend. A recent study conducted by the Nature Conservancy estimates that 21 percent of higher plant species could be lost if world temperatures rise even five degrees Fahrenheit early in the next century. Nearly half of the plant species federally classified as Threatened or Endangered could vanish entirely under these conditions, with flora in central Texas, the Florida Panhandle, and the southern Appalachians subject to the greatest risk.

Protection under the provisions of the federal Endangered Species Act is no guarantee that a plant species will recover and be taken off the doomsday list. Due to the fact that many plant species are often not listed until their individual numbers are dangerously low, prospects for recovery are much lower than they are for higher animals. Of the approximately 492 plants and animals currently receiving federal protection, the median number of surviving individual plants of a given species is a mere 120, as opposed to 1075 for endangered vertebrates and 999 for invertebrates. Many of the critically endangered plants species hang on in very small, fragmented habitats that are highly vulnerable to further, or even complete, destruction by fire, severe storms, or large-scale development.

Plants need not be as scarce as hen's teeth to be considered candidates for ecological disaster—entire forests have slipped into the danger zone over the past three decades. The industrialized world today pumps more than 100 million tons of sulphur dioxides into the global atmosphere via factory and auto emissions, and this toxic crud is already having a profound and detrimental effect upon the world's forests. It is estimated that more than half of Germany's forests are already severely affected, with the German fir very close to extinction. One in ten of the trees in Czechoslovakia and Sweden are dead and the forests of the southern Appalachians are under growing stress from air pollution and acid rain. Yet acid-producing emissions are expected to increase by one-third by the end of the century.

At present, no North American tree species are considered to be in imminent danger of extinction, but this will not be the case if the effects of acid precipitation and smog increase over the next few decades. The red spruce in Vermont is considered at risk due to acid precipitation, and some species might decline to the point of at least commercial extinction if clear-cutting is allowed to proceed unregulated in the ancient, old-growth forests of the Pacific Northwest. Botanist F. Schuyler Mathews, writing in 1915 of then rapacious and virtually uncontrolled logging on public lands, insisted that "dollars actually hold the woods and when there are enough of them passing from hand to hand, the trees will go." Things haven't changed much since then, for economics and personal gain still determine the fate of the forests of North America—and virtually everywhere else in the world—in the latter years of the 20th century.

Acid rain and air pollution are playing havoc with forests in many areas of the world. In North America, the area of the worst acid precipitation contamination and ozone pollution lies in a broad swath from southern Quebec and Ontario to Missouri and South Carolina. The areas of the highest concentrations of sulphates delivered via precipitation—at 31 pounds per acre—are western Pennsylvania, Ohio, Indiana, and parts of the southern Appalachian Mountains.

The sprawling Shenandoah National Park extends for nearly 100 miles along the spine of the Blue Ridge Mountains of Virginia. It is a place of exquisite, rugged natural beauty but it is also an ecosystem in grievous trouble. Sulphate levels in the park are estimated at about 22 pounds per acre. On a good day a visitor might scan mountain ranges 30 miles distant; on a bad one, a mere 10 miles. When the park was established in 1938, the view routinely took in vistas extending 90 miles, the clear air containing only invisible natural particulates. Today the ever-present smog ensures that summertime views from the ridges have declined by 80 percent over the past 50 years. But the quality of the view is far from the only worry. In the past 25 years, Shenandoah's air has become dense with sulphates, nitrogen oxide, and ozone that is slowly but surely destroying the very forests themselves, along with the many creatures that rely on them for food and shelter.

The principal sources of this pollution are auto emissions and the multitude of industrial plants that have sprung up in the surrounding region. Combined, they have created an air pollution problem that has placed the park in a state of environmental crisis exceeded only by that of the beleaguered Everglades in South Florida.

The smog damage to vegetation nationwide has been well documented over the past three decades, and the insidious destruction is especially severe in the Shenandoah region. A 1992 Pennsylvania State University study found that of 60 black cherry trees examined at a single location in the park, 52 showed evidence of ozone damage. More than half of 245 trees of several species surveyed in the park had moderate to severe ozone damage—the so-called "stipple effect" of leaf blistering.

Reforestation of denuded woodlands is an environmental concept rapidly gaining favor these days, though the timber industry has been doing it on a large scale for years. Some states administer reforestation programs and maintain tree farms for the purpose of propagating desirable tree species and distributing them to public agencies and residents who wish to revegetate their holdings. My home state of New Jersey operates a single, little-known tree farm in sprawling, rural Jackson Township, near the famous Pine Barrens. The 450-acre State Forest Tree Nursery, the last one in the state and one of the few in the entire northeast, annually distributes some 750,000 seedlings to private individuals and businesses statewide and seeds to many thousands more. The farm, located on the site of the state's old quail farm, produces both hardwood and softwood trees, including white, loblolly, and Japanese black pines; tulip trees; poplars; and black locust. The seedlings are offered at about half the price charged by private nurseries.

Biodiversity has already been lost in the food plants mankind depend on for survival. Whereas early man harvested and lived off several thousand plant species, today humanity depends on a mere 150 cultivated species, with most of the world relying on no more than 12 plant species for its daily bread. While those "amber waves of grain" are poetic and awesome in visual impact, such a monoculture presents an open invitation to insect pests, which then develop immunity to the pesticides that further damage and reduce the biodiversity of the earth in their indiscriminate use.

The survivors among the plants worldwide will include thousands of hardy and adaptable species able to exploit the altered and damaged habitats and lower air quality of tomorrow's urbanized environments. Among the grasses alone, hundreds of species readily colonize construction sites, power line rights-of-way, roadsides, and other such disturbed habitats, and will likely thrive in the coming century. Likewise, pest species such as the thistle, horse nettle, burdock, jimson weed, and cocklebur are everywhere humankind has left its mark on the land. Given such a large roster of botanical bad guys, one must draw the line somewhere. Therefore, I will discuss 21 of the more familiar and conspicuous aquatic and terrestrial plants that clearly exhibit those qualities that foster survivorship among the plant kingdom. For more in-depth information on plant conservation and the biology of weed species, I encourage you to refer to the appropriate titles in the bibliography.

Common reed
Phragmites autumnalis

This regal-looking plant, perhaps more so than any other, has come to symbolize environmental degradation in North America. In addition to being found everywhere that marginal lands have been altered by man, it is so large and showy that its presence cannot be overlooked. The plant has become almost synonymous with the industrialized marshlands of our cities.

The common reed, also called foxtail, plume grass, and, simply, phragmites, is a true plant camp follower of industrial mankind. It quickly and completely takes over tidal marshes that have become degraded due to pollution, filling, or diking, and forms vast urban prairies of virtually single-species vegetation with little food value to wildlife.

Phragmites is native to Eurasia, but through both accidental and deliberate transplants, it is now found virtually worldwide in disturbed habitats. The plant was brought to this country in the last century as an ornamental, because its tall (up to 20 feet) graceful stems and feathery seed heads present an exotic tropical appearance. The plant spreads by means of runners and underground shoots, and as the tiny seeds are also widely dispersed by the wind, the aggressive reed quickly made itself at home and began to conquer new territory. Although great areas of phragmites often catch fire and burn spectacularly out of control, the plant's roots are usually situated deep enough in the marsh mud to escape damage and fire has been shown to actually invigorate subsequent growth. Although the plant is now found from coast to coast, it is nowhere more apparent than in the broad New Jersey meadows just to the south of New York City.

The so-called "Meadowlands," a political rather than a biological entity, is a greviously man-handled saltmarsh that has become almost completely dominated by phragmites over the past century. With the exception of isolated remnant stands of native saltmarsh vegetation found here and there throughout this urbanized environmental Armageddon, the reed is the dominant plant of the developed meadows. This once great tidal estuary, the respective drainages of the Passaic and Hackensack Rivers, has been repeatedly diked, drained, and filled over the years, and its verdant banks smothered under a maze of industrial and transportation constructions of every size and description.

This grim habitat holds no fears for the common reed, however. It seems made to order for the plant, and a traveler on the New Jersey Turnpike where it passes through this region can get a first-hand look at survivorship among the plants. In summer the vast, green sward—an "urban Serengeti," in the words of one botanist—reaches from horizon to horizon. In winter the sere, golden-brown expanse serves as a backdrop for hordes of wintering gulls and crows that soar above the flat landscape enroute to the landfills that still edge their odorous way into the swamps.

The reed has shown itself to be marvelously adaptable, and might even prove to be helpful to mankind in the future. The ubiquitous rangy plant is the subject of a brand new approach to sewage treatment technology that holds great promise for ridding the western world of the increasing megatons of waste sludge—one of the more dangerous byproducts of industrial civilization.

In a sludge-disposal technique developed in Europe in the 1950s and refined by a New York company, stands of reeds are planted in large holding tanks containing a layer of gravel and a layer of sand. Treated liquid sludge is applied to the surface and the established plants feed on the sludge and

Phragmites

Quinn '93

stabilize it. The reeds are cut and composted each winter, and the sludge applications continue for a ten-year period. The tanks are then drained and emptied and the process begun again. About 60 such reed-sludge beds are in operation in 11 states at the present time. The stabilized and detoxified sludge is suitable for use as fertilizer or ground cover, the developers say.

The phragmites reed has been used for centuries as roof thatching in its native lands, though its only possible alternative use here in industrial America—aside from that of a showy room decoration—might be as a new source of cellulose. This option is currently being explored by several companies, who hope to take advantage of a crop that is easily grown in vast quantities on marginal lands in the shadow of virtually all of our major cities.

The notion of the reed as a human food source is one that likely never crosses the minds of those motorists traversing the reedy acreage of Northern New Jersey, but in fact parts of the reed are passable fare. The young, green shoots can be dried and pounded into a fine powder that has a pleasant taste when moistened and then roasted like marshmallows. The fleshy rootstocks can be washed and dried to yield flour.

Hydrilla and water hyacinth
Hydrilla verspicillata and *Eichornia crassiseps*

Water hyacinth

Hydrilla

These two aquatic plants are typical of those species that were initially introduced for their decorative charm but soon got away from controlled cultivation and invaded the natural environment.

The beautiful but immensely troublesome water hyacinth was deliberately introduced into the United States as an ornamental plant early in this century. Native to much of tropical South America and lacking natural enemies here, the hyacinth bloomed in abundance, and inevitably entered aquatic ecosystems near New Orleans. It soon spread rapidly throughout the low-lying marshes and canals of Louisiana, and is today found from Florida to Texas and beyond, wherever the climate is mild enough year-round to support the cold-intolerant plant.

The water hyacinth is an attractive plant with its large, round, furled leaves and bright purple flowers. It reproduces via runners, or plantlets, that sprout from the parent plant and eventually detach and float free. A single plant might replicate itself hundreds of times in a single growing season. The resulting mass, which can cover hundreds of acres of previously open water, consists of plant clusters nearly three feet high, supported by the characteristic globe-shaped stem floats. A lake or pond overcome by water hyacinth looks very much like a broad, green meadow sprinkled with purple flowers, but the illusion is quickly destroyed by the passage of a boat through the packed mass; slow, graceful ripples undulate through the greenery, giving away the true nature of the "meadow." The complex root networks are finely haired and dense, and offer shelter and spawning sites for smaller fish species, but they soon fill and choke the water space below, blocking sunlight and inhibiting the growth of other plants, smaller organisms, and, ultimately, fish life.

For the most part, the water hyacinth is controlled by chemical and mechanical means. The former involves spraying the massed invader with herbicides, a technique that gives many biologists misgivings. Weed-eating barges equipped with huge submerged blades prowl the infested southern canals and bayous, scooping up and masticating the hyacinth hordes into mulch as best they can. Both methods do a partial job at best, for in the warm plant-hospitable Gulf States the hyacinth always seems to keep one root-step ahead of its would-be destroyers.

Increasingly, control of the water hyacinth is being turned over to whatever natural enemies can be found in its native lands in a campaign of so-called biological control. As is often the case with foreign invaders, the cure must be imported as well. In the case of the hyacinth, this involves a South American species of weevil that dines on the hyacinth on its home turf.

So far, the weevil has worked—sort of—as a control agent; at Jean Lafitte National Park near New Orleans, the insect, which feeds on the plant's tissues, has reduced the hyacinth's numbers by about 30 or 40 percent, though the park still has to use the weed-eating boats to stay ahead of the fecund plant. The weevil has not yet created any control problems of its own in the alien ecosystem—a distinct peril that plagues all biological control programs. There is no "magic bullet" when it comes to putting the brakes on the spread of such an adaptable and hardy pest.

Abundant as this species is today, you might think the idea of using it for human food would be explored. The young, light-green leaves and stalks are delicious when boiled and served with butter, or crisp-fried. The flowers can also be boiled and eaten.

Hydrilla is a Eurasian water plant that escaped cultivation as a pond and aquarium decoration, and is spreading throughout the southeastern part of the country, where it chokes lakes and ponds and crowds out native vegetation. Another exotic organism, the strongly vegetarian grass carp (Ctenopharygodon idellus) has been propagated and released by a number of states to combat this pest plant.

Goldenrod
Solidago canadensis

There are some 100 species of goldenrod in the world, the majority of them found in North America, with at least 75 species endemic to the eastern United States. Goldenrods are common plants from the high-tide line of ocean beaches to power line rights-of-way and interstate freeway verges. Where the common goldenrod has become well established, entire fields might be filled with the beautiful, golden-yellow blooms, making this showy plant one of the more attractive weeds.

Contrary to popular opinion, goldenrod is not one of the major banes of the hayfever sufferer. Unlike ragweed, which is a big pollen producer, the goldenrod depends on insect pollination of its lovely blooms for propagation. Plants with brightly-colored flowers do not generally broadcast pollen; instead they use their floral display to attract insects, which pick up the pollen spores hidden in the depths of the flower. Goldenrod pollen has been found to cause hayfever under controlled laboratory conditions, but outdoors there's simply not enough of it to be a problem for snifflers and sneezers.

The various goldenrod species range in height from three to six feet and are among the more visible weed plants along roads and in fallow fields. Under favorable conditions the plant soon forms large patches and attracts many pollinating insects, including desirable ones such as butterflies and the honeybee. The goldenrod was much admired by the early English botanist John Tradescant, who called it *Virga aurea Virgine* when he found it in Virginia during a visit in the late 1600s. He subsequently brought it back to Britain where it has acclimatized too well, becoming at first a beloved garden plant and then a colorful but obnoxious weed, as it is mostly regarded here.

In spite of their undeserved reputation as hayfever pollen producers, goldenrod plants are fairly popular today as home decorations, and bunches of the flowering plants are hung in cool, dry places in the fall for use as winter decor. The sweet goldenrod (S. odora) yields a delicious tea when the fresh or dried leaves and flowers are steeped in boiling water for about 10 minutes.

Though the common goldenrod is hardy and abundant, one species of the family is considered threatened, at least in New York and Michigan. The Houghton's goldenrod (S. houghtonii) occurs in the region of the Great Lakes, living in the specialized dune habitats found along the northern shores of Lakes Michigan and Huron. In New York the plant occurs in a single colony in a meadow in Genessee County. This goldenrod is named for its discoverer, Douglass Houghton, a Michigan state geologist in the early 1800s.

Milkweed
Asclepias syriaca

The warty, gourd-shaped pods of the common milkweed are among the more familiar winter sights in disturbed and waste habitats everywhere. The summer plant is recognizable by its lavender to purple flower cluster and fleshy, oval shaped leaves. The sticky, milky fluid that emerges from a break when a leaf is removed is a familiar characteristic.

Milkweed is the primary food plant of the monarch butterfly. It is considered an edible plant because the young leaves and pods as well as the flower buds can be eaten by people if they are boiled, drained, and boiled several times again in order to purge the plant's bitter, mildly toxic juice. The first few water changes should be rapid—about every minute—using boiling water to make the changes (cold water tends to fix their bitter flavor). The mature flowers of the plant can be submerged in boiling water for about a minute, dipped in batter and fried.

At the other end of the scale from the ubiquitous common milkweed is the rare and endangered Mead's milkweed (A. meadii). This plant, found in scattered colonies in northern Illinois, Iowa, Wisconsin, Indiana, and Kansas, was apparently never very abundant. It is thought that as settlers moved westward in the last century, the plowing and tilling of the prairies severely reduced this plant's numbers. Today the Mead's milkweed survives in isolated colonies—some consisting of only a few plants—in Illinois, Missouri, and Kansas. It is presently being considered for listing as an endangered species.

Common mullein
Verbascum thapsus

The mullein is another Eurasian import that has made the best of the altered landscapes offered by modern industrial America. This hardy plant is found wherever there is enough soil to support it, and is often the most common and noticeable vegetation in gravel pits, railroad rights-of-way, construction sites, landfills, and other such habitats. It is highly tolerant of smog and contaminated soils, and can survive with ease in grim environments such as oil tank farms, dredge spoil banks, and slag heaps.

Mullein is one of the easiest plants to identify; it is a rather coarse-looking plant and at the same time has an air of elegance about it. The light green leaves gathered in a dense rosette on the ground are thick, feltlike, and hairy, and the tall, central stem topped by the wandlike fruiting head remind you of the bizarre and exotic Century plant of South America. Mullein is a biennial, producing the fleshy leaves in its first year and the flowering spike in the second. The spike remains standing long after the onset of cold weather, and is a prominent winter feature of the barren landscapes this plant calls home.

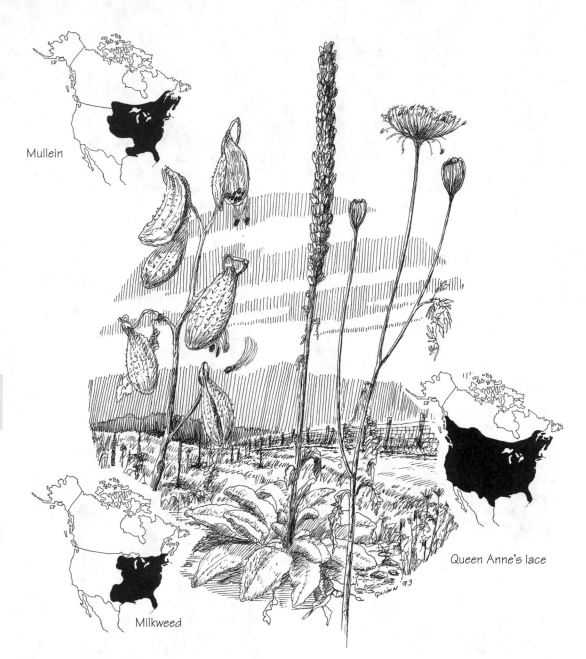

Mullein

Queen Anne's lace

Milkweed

Steeped in boiling water for about 10 minutes, the dried leaves of the mullein plant yield a perfectly acceptable tea.

Queen Anne's lace
Daucus carota

This ubiquitous plant, sometimes called wild carrot, grows between one and five feet high. Like the domesticated garden carrot, it is a member of the parsley family (*Umbelliferae*). Queen Anne's lace is a cosmopolitan species, occurring throughout temperate North America, in the northern British

Isles, and in Europe from Scandinavia to the Mediterranean. In North America it flowers from May to October, and its beautiful and characteristic tiny white flowers arranged into lacy compound "umbels" do have a "lace doily" look to them. The single, small, central flower, sometimes hard to spot amidst the complex white mass, is usually a dark red or purple color. After flowering, the drying stalks of the umbel begin to fold upward into a familiar fall shape called a "bird's nest."

The wild carrot produces an average of 4000 seeds per plant, a fact that no doubt has something to do with its continued abundance and ability to spread. It is a biennial; rosettes of old leaves lie flat on the ground throughout the winter, sending up new plants the next spring.

Queens Anne's lace is an edible plant; the large, first-year tuber can be prepared exactly as you would the domesticated carrot.

Ragweed
Ambrosia artemisiifolia

The generic name of this weed species means "delightful to smell," though millions of hayfever sufferers might seriously question that attribute! The venerable Gray's Manual of Botany refers to this ubiquitous plant as "a despised weed," no doubt a label much closer to the hearts—or sinuses—of August-to-October snufflers everywhere!

Two species of ragweed occur throughout most of North America: common ragweed (A. artemisiifolia) and great ragweed (A. trifida). They are similar in appearance—tall, rangy plants with alternately emerging, upward curving stems—but the great ragweed reaches a height of 15 feet, about twice that of the common ragweed, and is much coarser and more bristly in stem texture. The flowering period is from July to October.

Ragweed is native to much of North America, most abundant in the east. It has been introduced into Britain, where it has become locally acclimatized and is called Roman wormwood. It is the classic invader species of the biologists, one of the first coarse plants to move in and colonize habitats that have been burned, blighted, or bulldozed into oblivion.

Japanese knotweed
Polyganum cuspidatum

Also known as American or Mexican bamboo and wild rhubarb, this introduced exotic is a common pest species and the bane of many a suburban gardener. Of Asian origin, knotweed is another of those plant species that was introduced with the best intentions—it seemed like a harmless and attractive ornamental plant where it belonged back home, but in the New World it took off like wildfire and soon wore out its welcome.

Japanese knotweed spreads by means of underground rhizomes; the entire root system must be dug up to thoroughly evict the plant from the premises. The plant grows into great dense thickets up to nine or ten feet tall, and the green-and-red-spotted, hollow, bamboolike stems do lend an exotic tropical appearance to the temperate suburban garden.

Japanese knotweed is edible and, in keeping with one of its common names, does indeed taste like rhubarb when properly prepared. The very young spotted shoots should be picked as soon as they emerge from the spring soil. Boiled in sugared water and served with sugar, they have a delightfully tart, rhubarby taste. The young shoots can also be collected when they are between six and ten inches tall, steamed for three or four minutes, and served as an asparagus substitute. They can also be cooked with cranberries to prepare a delicious cold sauce. Older stems can be peeled and the peels boiled with sugar and pectin to yield a tasty jam.

Pokeweed
Phytolacca americana

Pokeweed is common in disturbed areas and quickly moves in when the endemic vegetation has been removed. It is a highly conspicuous plant, almost treelike in size (to over six feet) and appearance. The other common name, inkberry, is due to its dark red berries, which are sometimes crushed and made into a drawing ink by children.

Pokeweed is a poisonous, though still edible plant. The large taproot, seeds, and mature leaves are dangerously toxic if eaten raw in any quantity. The very young plants are sometimes boiled in two changes of water and served as salad greens with no ill effect, but in today's supermarket world few are willing to take the chance of eating a possibly poisonous plant, or even to seek it out in the isolated areas, such as railroad rights-of-way and abandoned landfills, in which it is often found in abundance. The plant flowers from July to September, when the large, conspicuous blue-black berries appear.

Poke is found throughout much of eastern North America and southern Canada, and has been introduced into Great Britain, where it has become locally acclimatized.

Crabgrass and other "weeds"
Digitaria sanguinalis

The grasses (Gramineae) are among the most successful and hardy of the plant survivors. They are self-pollinating; the wind and the appetites of birds rapidly scatter their seeds everywhere. Worldwide, there are about 600 grass genera comprising about 10,000 known species. Some of them, such as the indispensable oats, wheat, rye, and rice, are most assuredly friends of man, while others have less worth and thus generally qualify simply as weeds. Crabgrass is native to southern Europe, and is thought to have arrived in both Britain and the United States via shipments of impure grain in the late 18th century. The plant has also migrated to many parts of South America, probably by the same means, for few would consider crabgrass any sort of ornamental worth transplanting deliberately.

Crabgrass has five flowering spikelets that radiate from a single stalk like fingers, hence its common name in England, "hairy finger grass." A similar, closely related species, smooth crabgrass (D. ischaemum), is native to Eurasia but has also found its way here. Another ubiquitous lawn pest, it can now be found throughout the United States with the exception of southern Florida and parts of the arid southwest.

Witch grass (Agropyron repens) is another lawn weed of undisputed pest status. A native of Europe, it has emigrated by unknown means to the British Isles, Canada, and all of the United States except for the extreme South. It is a highly persistent plant, rapidly spreading by means of both abundant seeds and tenacious rootstock, which can produce a new plant anywhere along its considerable length.

Yard grass, or silver crabgrass (Eleusine indica) is a dark greenish weed naturalized from southern Europe. It resembles crabgrass but has denser flower spikelets and grows in thicker tufts in lawns, where it is universally regarded as a coarse nuisance plant. It is found throughout the United States except for northern Maine and parts of the north-central states.

Depending on where it's found, meadow grass (Poa pratensis) can be considered a blessing or a curse. Known also as Kentucky bluegrass, this species is the principal and desired ingredient of beautiful lawns everywhere, but where it becomes established in gardens and waste areas, it grows much more uninhibitedly (from 4 to 36 inches in height) and is summarily dismissed as a weed pest. No one knows for sure just where this plant originated, but today meadow grass is found in abundance throughout North America, temperate Asia, North Africa, and most of Europe, where it is primarily found at higher altitudes in mountainous regions.

The very closely related spear grass, or annual bluegrass (Poa annua) is a smaller species, seldom attaining a height of more than 12 inches. It is an amazingly adaptable plant, and is the grass most likely to be seen growing in such unorthodox places as cracks in highway pavements, roof gutters, window ledges, or other places where a bit of soil or pockets of organic debris might collect and a root can gain a toehold. A native of Eurasia, spear grass is now found virtually worldwide, even in the tropics where it usually occupies higher altitude montane environments.

White and red clover
Trifolium repens and pratense

The clovers (Papilionaceae) are among the commonest and best-known lawn and field plants in the world. The family name stems from the French, papilon, or butterfly, from the flower heads' fancied resemblance to these colorful insects. The white and red species are the two most familiar and widespread species, but the crimson clover (T. incarnatum) and the zig-zag clover (T. medium), both common weeds, are distinguished primarily by subtle differences in leaf pattern and flower color differences. This is the province of botanists and for our purposes, simply dividing them into the red and white varieties will do.

Clovers

White or Dutch clover, the more common species on suburban lawns, is recognized by the color of its rather small flowers—white tinged underneath with pink—and by the dark patch bordered by a white band on each rounded leaflet. This clover seldom exceeds the height of six inches, and is thus favored by lawn fanatics over the taller, bigger-leaved, and rangier red clover. White clover is highly attractive to pollinating insects; a lawn filled with this plant will often swarm with honeybees on a warm summer day.

Red clover has a much larger bloom—pinkish red or pinkish purple—and its three-pointed leaves are more elongate in outline, dark green, and crossed by a triangular whitish band. The red clover is found throughout continental Europe, the British Isles, and North and South America. Although this clover occurs in unmanicured lawns and in cemeteries, parks, and unmowed highway verges, it is much more abundant in pastures, along secondary roads, and in overgrown waste areas.

The dried flowerheads and seeds of both species of clover can be ground and rendered into a nutritious flour. The flowerheads also make a healthful tea, especially if mixed as an accent to other teas.

Dandelion
taraxacum officinale

This amazing weed is best known to North Americans as a persistent invader of otherwise perfect lawns, and as the principal ingredient of that homemade wine made famous by fantasy writer Ray

Dandelion

Bradbury. The dandelion's common name stems from the French *dent de lion*, meaning "tooth of the lion," a reference to the plant's toothlike, lobed leaves. The hardy weed is at once despised and loved by humankind; it is a singular source of irritation to the meticulous gardener seeking to evict it from lawn or rock garden, and a symbol of hope and cheer amid the grays of November, when its sprightly yellow flowers lend the last vestige of summer color to the scruffy, browning lawns of the dying year.

The dandelion is a native of Eurasia that is now found around the world in the northern hemisphere. In addition to its preferred habitat—the carefully tended suburban lawn—the plant demonstrates an amazing ability to survive in just about any situation, even between the pavement cracks of busy superhighways, where fully adult, flowering individuals can often be found keeping a very low profile indeed.

Although, with the exception of dandelion wine-makers and natural salad lovers, this plant is almost universally considered a pest in this country, this is not the case in Europe. There, the dandelion has long been appreciated for its health and medicinal benefits. Its leaves and blossoms contain more vitamin C and A than most garden vegetables, and are a healthful stimulant to various internal organs, especially the kidneys and bladder. In this respect the dandelion has long been called "potty herb," "pee-a-bed," and "wet-a-bed" in Britain and *pissenlit* in France.

The high food value of the lowly dandelion can be readily appreciated in a comparison of one pound of the plant with an equal amount of garden lettuce, fresh and uncooked:

	Dandelion	Lettuce
Protein (grams)	12.3	3.8
Fat (grams)	3.2	0.6
Carbohydrates (grams)	40.0	9.1
Calcium (mgs.)	849.	194.
Phosphorus (mgs.)	318.	63.
Vitamin A (I.U.)	61,970.	5060.
Iron (mgs.)	14.	3.4
Thiamin (mgs.)	0.85	0.14
Riboflavin (mgs.)	0.65	0.26
Niacin (mgs.)	3.8	0.6
Vitamin C (mgs.)	163.0	57.0

On the minus side, the deep-rooted dandelion absorbs about three times as much iron, copper, and other nutrients from the soil as most other plants, and exudes ethylene gas, which inhibits the growth of the more desirable plants that share the environment with it. If dandelions are removed from a lawn or garden and composted, however, these stolen riches are recycled into the soil where the more shallow-rooted garden flowers and vegetables can make use of them.

Practical considerations aside, the featureless lawns and waysides of today's urbanized landscapes would be considerably less colorful places were the perky little dandelion ever to be completely exterminated. Its round fluffy seedheads, scattered over the summer lawns of suburbia and filled with tiny parachutes just begging to be huffed away are still an irresistible lure to eager kids.

Plantain
Plantago major

There are four species of introduced plantain, all of them Eurasian in origin and all imported by accident into the United States and Canada, where they are familiar weeds in lawns, gardens, public parks, and waste areas. The common or broad-leaf plantain and the similar black-seed plantain (*P. rugelii*) both have oval, pointed, strongly ribbed leaves that lie rather flat on the ground in a rosette pattern. The seed stalk of the common plantain is thick and dense with seeds, while that of the black-seed variety is slender and sparsely flowered.

These plants are found throughout North America. Both flower from June to October.

The English or buckhorn plantain, or ribgrass (*P. lanceolata*) is a European weed that has become established throughout the United States and Canada. It occurs naturally throughout Europe, as far north as Iceland. It is distinguished from the other plantains by its long, narrow lancelike leaves with prominent ribs, or veins, and by the single flower knob with flowers protruding from it like a pincushion.

Plantain

The hoary plantain, or lamb's tongue (*P. media*), is yet another European import now found in local abundance in the eastern half of North America. The flower spike is stubby, much like the buckhorn plantain, but is slightly more elongated, and the flowers are light purple rather than whitish.

The name plantain all too often brings to mind aggressive coarse weeds that must be rooted and poisoned from lawns and gardens, but not all plantains are noxious botanical immigrants. The heart-leaved plantain (*P. crudata*) is a bona fide native plant, and is considered a candidate for inclusion on the federal endangered species list. This attractive plantain occupies shaded streambank habitats from New York to Minnesota. After 1935 the species began a dramatic decline, and by 1968 researchers were able to locate only a few scattered colonies of the plant in Ontario, New York, North Carolina, Illinois, and Missouri. This plant seems to favor semiaquatic habitats, even growing in shallow, flowing waters, and probably the damming and alteration of streams everywhere has been one of the principal causes of its gradual disappearance.

If collected when very young, the leaves of the common plantain are a delicious addition to any salad or can be boiled and served with butter; older leaves become too tough and stringy for culinary use.

Plane tree
Platanus occidentalis

The plane tree is also known as the sycamore and the buttonwood tree. The Oriental sycamore (*P. orientalis*) is an introduced exotic, widely planted as an ornamental tree in the early years of this century and now widespread throughout eastern North America. The California sycamore (*P. racemosa*) is an abundant tree of the coast ranges and interior valleys.

Plane Tree

This large and attractive tree is one of the relatively few foreigners among the plants that has not worn out its welcome, being considered a decorative asset to city streets throughout its range. The tree is highly tolerant of salt and air pollution and the general wear-and-tear of urban life, and is still widely planted as an ornamental.

The plane tree reaches heights between 60 and 80 feet. The main trunk is deeply furrowed and lumpy near the base, and the bark peels off in large flakes and patches that leave the buff-colored inner bark exposed. The resulting mottled and variegated appearance is the best fieldmark for this species. The upper branches of the tree are often a chalky white, lending a striking, exotic look in winter to residential streets lined with larger specimens.

Although the plane tree has no commercial value today, the strong, coarse-grained, heavy wood was once used in the manufacture of ox yokes, cigar boxes, furniture, and interior finishes.

Willow
Salix babylonica

The willows are a very large family, with at least 32 species and various hybrids between them existing in North America. All willows are water-loving trees found primarily near rivers and lakes. As these habitats are altered or destroyed many of the native species will become increasingly rare and eventually threatened. The weeping willow, an introduced exotic native to China, is a large graceful tree widely planted as an ornamental. It was introduced into this country during colonial times and has spread throughout the eastern United States, usually following river courses, where it propagates through the rooting of detached branches.

The willow is a medium-sized tree, attaining a height of 40 to 50 feet with a graceful, rounded crown that looks like a green fountain. The tree grows rapidly but is relatively short-lived, a fact that might comfort homeowners whose septic tanks and sewer pipes have been invaded and clogged by the water-seeking root systems.

The crack willow (S. fragilis) is another large naturalized European and Asian species that was planted as an ornamental tree about 100 years ago in the vicinity of Boston and other New England cities. It is now widespread from Quebec to Pennsylvania and Georgia.

The widespread and familiar pussy willow (S. discolor) is a native plant that has nonetheless been widely transplanted outside the natural range as a garden ornamental. It is found from southern Canada south to Missouri and Virginia, always on low, wet ground.

Ailanthus
Ailanthus glandulosa

This attractive, under-appreciated tree is an import from China, where it has long been known as "Tree of Heaven" and "Paradise Tree." In North America it is generally regarded as an attractive weed species. The ailanthus was brought to this country for the first time in 1784 and later, from Europe, by one William Prince, who made numerous plantings of the tree in the area of Flushing, New York in 1820. The species spread quickly and attained such prominence and familiarity as a "city tree" that it was the botanical "star" of the classic movie, A Tree Grows in Brooklyn.

Ailanthus attains a maximum height of about 75 feet, with 40 or 50 feet a more common average. At maturity it is an attractive tree, its large, compound, palmlike leaves lending it a vaguely tropical look. The species has become naturalized throughout the eastern United States and southern Canada.

Ailanthus is a very rapid grower, spreading quickly by root suckers and colonizing all manner of hostile urban habitats, even those that cannot support any but the very hardiest of weed flora. Where it invades suburban and outlying areas it competes with more valuable tree species, often crowding them out. It is resistant to air pollution and able to survive under the most adverse urban conditions.

Large specimens of ailanthus can have a trunk diameter of two to five feet. The wood is soft and brittle and of no commercial value; commercial logging is one threat to survival this hardy tree does not face.

Poison ivy
Rhus toxicodendron

This plant, and the closely related poison oaks and sumacs, are members of the tropical cashew family. It is widespread throughout most of the United States and southern Canada, inhabiting all types of natural and man-altered habitats, and favoring disturbed areas, such as roadsides,

Ailanthus

fencerows, and construction areas. Continuous human tinkering with natural habitats has doubtless aided the rapid spread of this nuisance plant.

Poison ivy is distinguished by its compound groups of three shiny oval leaves, and compact clusters of whitish flowers and berries. The plant thrives in wet or dry situations, though it does best when at least partially shaded.

All parts of the poison ivy plant are poisonous; how poisonous depends on your individual sensitivity to it. Poison ivy is a rangy, woody shrub that can attain considerable height; I am familiar with one impressive plant growing on the bank of the Shark River estuary on the New Jersey shore, reaching a height of about 30 feet. This particular plant, forming a dense tangled thicket, uses a large red cedar as a climbing base; the poisonous thicket has long been used as an outdoor lavatory by local anglers, who are probably unaware of the plant's nefarious identity!

The closely-related poison oak (*R. quercifolia*) is a small shrub (12 to 20 inches high) found in waste and disturbed areas from Virginia south to Florida and westward.

Poison sumac (*R. vernix*) can attain the height of 20 feet in the South, about half that in the northern parts of the range. It is found from Maine west to Minnesota and south to Louisiana and

Poison ivy

Florida. All parts of this plant are intensely poisonous, the acid-oil toxin working quickly and often painfully on contact with the skin. Poison sumac is always found in low, wet situations and in swamps.

Staghorn sumac
Rhus typhina

A very common and familiar plant, staghorn sumac occurs throughout eastern North America west to Minnesota and Missouri in temperate regions, and is abundant throughout New England. This sumac will grow in a wide variety of soils and situations, and is adept at colonizing abandoned landfills, construction sites, and the like. It is a familiar roadside plant, and rivals the Ailanthus in its

Staghorn sumac

35

ability to invade urban areas. The plant's common name stems from the conspicuous fruit, which takes the form of deep-red, conical clusters of fuzzy berries said to resemble the antlers of deer, or stags, in velvet. The fruiting bodies are an important winter food staple of many bird species.

Sumac can reach heights of up to 40 feet and trunk diameters to 10 inches, though most plants are about half that size, resembling rangy shrubs. The compound leaves turn bright scarlet in the fall, lending a touch of color to many an otherwise undistinguished roadside.

Although the light, brittle wood of this plant has little use today, it was once used in the manufacture of cabinetry and maple sugar sap taps, or spiles. The ripe berries of the staghorn sumac can be rendered into a delightful cold drink. The entire fruit cluster should be collected and rubbed to bruise the berries. Soak the berries in cold water for about 15 minutes and then pour the pinkish water through a sieve to remove any debris and hairs. Chilled and sweetened to taste, the beverage tastes rather like pink lemonade.

Kudzu
Pueraria lobata

The rapid spread of this showy but immensely troublesome vine is a classic example of the old "too much of a good thing" axiom. Native to Japan and parts of the Asian mainland, the growth and nitrogen-producing capabilities of kudzu were noted in the plant's natural habitats, and it was brought to the United States in 1876 as a decorative plant, erosion control agent, and a means of returning nitrogen to depleted soils. The vine soon escaped cultivation and has today spread throughout the southeastern states, from South Florida to northern Maryland.

A strongly invasive plant, kudzu generates tough, lengthy fronds that cover trees, telephone poles, buildings, and any other objects that offer it support, quickly forming dense green cloaks that deprive other plants of critical sunlight. It can grow a foot a day under favorable conditions, and demonstrates accelerated growth rates in response to higher carbon dioxide levels such as those in city environments. Kudzu has generally made life difficult for highway maintenance crews throughout the Deep South; it must be regularly cleared from highway shoulders and directional signs.

Although kudzu moved northward at a rather modest rate at first, scientists now fear that it will soon accelerate its invasion of new territory if predictions on global warming prove correct. The plant is not overly cold-tolerant, a factor that has so far limited its northward expansion. If global temperatures rise and atmospheric carbon dioxide levels double, however, researchers think that kudzu will respond by growing at least 50 percent faster and denser, and moving north at a much greater rate, possibly reaching the Great Lakes region by the year 2030.

Kudzu is a plant better adapted to the takeover of agricultural and natural areas, and thus it is not expected to become a real threat in the heavily urbanized North, even if it does arrive there full of fight early in the next century.

Yet another potentially serious weed pest already on the march throughout much of the southeast is the Japanese climbing fern (Lygodium japonicum). This rather attractive, lacy-leaved vine, first imported into North Carolina around 1900 as an ornamental plant, has become established in shaded wetland areas, such as flood plain forests, in parts of Florida, the Gulf States, and Texas. Like kudzu, this fern is a fast-growing vine that forms dense tangles of 6- to 100-foot fronds that deprive other plants of necessary sunlight. Biologists fear that the threat of a real takeover is highest in South Florida, where the year-round mild temperatures are more conducive to nonstop growth.

Kudzu

Japanese climbing fern

Chapter three

The insects

The sense of death is most in apprehension
And the poor beetle, that we tread upon,
In corporal sufferance finds a pang as great
As when a giant dies . . .

William Shakespeare, *Measure for Measure*

Roughly 30 million insect species—more than 350,000 of beetles alone—share this world with us. That's species, not individuals. Individually, insects number in the countless trillions—as in grains of sand or blizzards of stars in the night sky. Entomologist Frank E. Lutz wrote in 1918 that there were more than 15,000 known species of insects to be found within 50 miles of New York City alone, 2000 of them butterflies and moths. A single swarm of locusts might weigh about 20,000 tons, and since an individual locust weighs about two and a half grams, the swarm might consist of about eight thousand million insects. The primitive, flealike little springtails, rarely noticed, might exist in a single acre of soil in densities close to one billion bugs.

Entomologist Brian Hocking, computing his datum on the average weight of an insect being 2.5 milligrams, speculates that the total weight of the world's living insects (numbering about 10^{18}) exceeds that of the total number of living men by a factor of 12.

Insects, even more than the essentially unloved reptiles and amphibians, enjoy a love-hate relationship with humankind. Naturalist Roger Tory Peterson, in his introduction to *A Field Guide to the Insects*, noted that people can be divided into two principal categories with regard to the insect kingdom: ". . . those who find insects endlessly fascinating and those who would like to get rid of them." Dr. Peterson didn't offer percentages for either camp, but I suspect the "bug-haters" far outnumber those who discern a cosmic beauty in the physical complexities of a mosquito, a housefly, or a cockroach.

Regardless of how you might react to the sight or touch of an insect, there is no doubt that a bugless environment is a desperately sick one, ecologically speaking. Although it's still difficult to spend more than 15 minutes outdoors almost anywhere during the warmer months today without interacting with one insect species or another, the diversity of insects, like that of all other living things, seems to be on the decline.

Despite the vaunted persistence of some of our more notorious insect pests, the truth of the matter is that insects as a group are highly vulnerable to changes in the habitat and to the introduction of toxins and pollutants. High levels of ozone and chlorinated hydrocarbons from industrial and auto emissions near our larger cities have greatly reduced the populations of both pest and beneficial insects over the past three decades. People comment wistfully on the scarcity of summer evening fireflies and luna moths, but they are far from the only insect casualties of urban America.

Toxic chemicals and organic pesticides have long been the heavy artillery in humankind's war with insects. One of the earliest was the highly toxic copper-arsenic compound, Paris green. This effective bug killer was developed in response to the eastward migration of the Colorado potato beetle between 1870 and 1890; through that period, the insects moved eastward from their native Rocky Mountain home at the rate of about 85 miles per day and by 1890 had made serious inroads into Maine's potato harvest.

Prior to 1900, however, most of the commonly employed insecticides were rather innocuous compounds utilizing natural materials. They included such exotic-sounding home remedies as whale oil soaps, pyrethrum, white lead, cayenne pepper, tobacco, hellebore, and borax. By the 1920s, the accepted potato bug destroyer was calcium arsenate, a deadly poison casually applied by hand in those innocent, pre-consumer-protection days.

Following World War II, the first broad-spectrum, synthetic pesticides came into being, with the notorious DDT at the vanguard. This chlorinated hydrocarbon compound was first synthesized in 1874 but it was not until 1939 that its application as an insect killer was fully appreciated and put to use.

The environmental devastation that spread across the land in the wake of DDT's heavy-handed use against the insect pests is well known, documented with force and poetry by biologist Rachel Carson in her landmark indictment of the pesticide industry, Silent Spring. The specter of a pesticide-ravaged biosphere emptied of its wonderful variety of life emerged and was noted by all who cherish the earth and its living things, but the end result of the indiscriminate spraying was something few had bargained for.

By 1960 over 137 insect species had developed genetic resistance to DDT and other chlorinated hydrocarbons, compared to a mere 20 species that developed resistance to the earlier, more primitive pesticides. By 1975 more than 300 pest-insect species were found to be DDT-resistant, and although application of the chemicals had increased from the 1951 level of 464,000 pounds to an astounding 1.4 billion pounds in 1971, crop losses due to insect damage nearly doubled.

The legacy of mankind's tampering with natural ecosystems through the heavy and unrelenting use of lethal chemicals is a grim one. Residues of toxic chlorinated hydrocarbons are still present everywhere—in the soil, water, and in the bodies of living beings. Even penguins as far away from the point of application as Antarctica have tested positive for toxic residues.

But the insects are superb survivors; they practically invented the word, as any agricultural entomologist will readily admit. It is no coincidence that many of the insect survivors of tomorrow are species that are or could be considered harmful to man—some of them ominously so.

The insects are grouped, even by laymen, into two supergroups: those that constitute a threat to human health, comfort, and welfare; and those that are considered beneficial and attractive to the eye. Cockroaches, bedbugs, lice, mosquitoes, houseflies, fleas, termites, and most ants are considered pests in that they directly assault our persons, food supplies, or dwellings. Potato beetles, weevils, tent caterpillars, gypsy moths, and cutworms wreak havoc on our crops and forests; they, too, are among "the enemy."

Butterflies, on the other hand, are almost universally admired and loved, though their fat, bristly, warty, and generally repulsive larvae might inspire loathing on the part of the average person. We prefer to disassociate the lordly and graceful tiger swallowtail from the obese, tube-shaped, bright-green caterpillar helping itself to our garden produce. But butterflies and moths, along with honeybees, ladybugs, praying mantids, fruit flies, and possibly dragonflies, are among the "good guys" of insectdom. Their activities and appetites directly or indirectly benefit us, or they simply lend action and color to the landscape.

Today, with easy-to-apply garden insecticides and electronic bug-zappers, the insects as an individual problem are becoming less evident. Wasps or june bugs on the screen, evening mosquitoes, and daylight-active biting flies are no longer the formidable summer foes they have been for generations of city and town dwellers. There just don't seem to be as many around any more. Fireflies, the "lightning bugs" chased and collected in jars by countless urban and suburban children of generations past, still flash their brilliant signals over rural fields and woodlands, but they've nearly vanished in significant numbers from the backyards and golf courses of suburbia.

House flies, however, are everywhere, and roaches will be happy and thriving, along with the rats, in the frozen and blasted wastes of the post-nuclear winter, remember?

There are some 90,000 species of insects inhabiting the United States and Canada. A good percentage of these will be the survivors of tomorrow to one degree or another. As long as a viable habitat remains to supply food and shelter, and mates can be located, many insect species will hang on for a long time. I have attempted, through the selection of a small number of the more visible hardy survivors of the present day, to present species that we can predict with reasonable certainty will persevere in the altered, fragmented, and polluted environments of the coming century.

German cockroach
Blattela germanica

The amazing cockroach is found throughout North America, not to mention most of the rest of the world; it is missing only in the coldest regions because it is a warmth-loving animal. Cockroaches, like rats and mice, are close associates of humankind, and are found in any moist, humid environment where food particles are likely to be found. Under laboratory conditions it has been found that cockroaches cannot long survive temperatures above 90°F and develop slowly at temperatures below 70°F. These preferences explain the insect's close association with people and their heated dwellings.

The German cockroach is one of the most common domestic cockroaches. Kitchens and bathrooms suit the roach's comfort-level nicely, but cockroaches can be found anywhere in a house; they live up to their official definition as "cosmopolitan domiciliary pests." The German cockroach was originally known to New Yorkers as the "Croton bug," due to its supposed entry into the city via the huge pipes that delivered water from the upstate Croton reservoir (it should be noted that the Greek word kroton means "tick or bug"). The English refer to the German cockroach as the "steam bug" or "shiner," the Spanish as la cucaracha.

The cockroach species are almost universally regarded with intense loathing, though they cannot truly be considered injurious insects. When compared with such invertebrate nemeses as locusts, lice, and malarial mosquitoes, the roaches are little more than ubiquitous nuisances that do little direct damage to home and person. Researchers who have worked with these fascinating insects speculate that much of the human hostility toward cockroaches stems from the perception of their presence in a home as a sign of uncleanliness. The creature's behavior also inspires hostility. The cockroach's fleetness of foot—up to three mph under controlled conditions—as well as its erratic movement and habit of sweeping its long antennae about in an intent, investigatory manner all have something to do with our deep-seated feelings of dread and disgust.

German cockroach

Roach mite

There are some 3500 described cockroach species worldwide, most of them living in the humid tropics. The family *Blattodea* are fairly evenly distributed over all of the major landmasses of the earth. The largest roaches can reach three inches in length and have wingspans in excess of seven inches; the smallest are hardly as large as a grain of rice. Cockroaches were probably introduced into the New World via African slave ships or European supply vessels servicing the Colonies. Today, cockroaches are regularly found in the luggage and on the clothing of ship and aircraft travelers, further attesting to the ease with which these insects are transported to fresh territories.

Other foreign interlopers among the roaches include the Oriental cockroach (*Blatta orientalis*), known in Britain as the "black clock" or "black beetle;" the smoky-brown roach (*Periplanata fulginosa*), common throughout the South; and the brown-banded cockroach (*Supella supellectilium*), a small, active roach that made its American debut in Key West in 1903 and spread across the continent by 1930. This roach can live all over a building, often laying its eggs in television sets, closets, and other warm dark places. The Mediterranean, or tawny cockroach (*Ectobius pallidus*) was first found in a summer cottage on Cape Cod in 1948. From there it quickly moved to the cities of Falmouth and Plymouth, Massachusetts, and has since spread throughout New England and much of the Northeast.

Cockroaches cannot bite, though the spines in their legs may cause irritation if the animal is handled. Except under the most filthy conditions, the insects themselves do not spread disease, nor has it ever been proven that they have played any role in human pathology. In fact, there is evidence that they secrete bactericidal substances from their tarsal pads that actually prevent the transmission of pathogens.

Captain Bligh of the infamous HMS *Bounty* reported that the numerous roaches in residence aboard ship were regularly treated to buckets of boiling water in attempts to control their numbers. A more recent (1961) report describes as many as 20,000 roaches killed in a single stateroom of a ship. The creatures had committed the usual roach sins against man by eating and fouling the crew's food, but they also reportedly devoured the leather of boots as well as the skin and nails of sleeping passengers.

Among some 50 species of cockroaches native to the United States is the so-called death's head cockroach (*Blaberus crannifer*), occurring throughout tropical America and native to the U.S. only in Key West, Florida. This is a large (50 mm), wild cockroach living as a scavenger amidst litter on the forest floor. This animal requires warmth, and although it has been raised as an experimental animal and shipped between points south and north, it has not yet become established outside its natural range. Special permits are required for its collection and rearing in captivity.

The desert cockroach (*Arenivaga bolliniana*) is one of several species of this genera found throughout the southwest and Mexico. This one lives in the nests of woodrats and scavenges decaying plant and animal material.

Cockroaches have long been valuable research animals, particularly in the areas of cancer, heart disease, and nutritional research, due to their hardiness and high reproductive rate. A hatchling roach develops into a near-perfect miniature of the adult within a half-hour and can begin to breed within a day.

In addition to mankind, the roach has many enemies. Chief among these are toads, some species of birds, spiders, including the hefty tarantulas, and, perhaps the most deadly of all, the cockroach mite. *Pimeliaphilis podapolipophagus*, a microscopic little monster with a scientific name longer than the largest American cockroach, can attack the roach in considerable numbers. A large German cockroach in a laboratory roach colony succumbed in an hour under the combined assault of 25 of these persistent little mites.

Perhaps the oddest instance of bizarre roach predation is a parasitic wasp. This small tropical wasp immobilizes the big jungle cockroach by inflicting a sting that doesn't kill the roach, but prevents it from moving unless it is led. The wasp leads the huge, now-compliant roach by its antenna to an underground lair, where the parasite lays an egg on the catatonic roach. The completely immobilized prey then serves as living food for the wasp's emerging larva. Ugh!

Silverfish
Lepisma saccharina

The strange, pallid little silverfish are members of the *Thysanura*, an ancient group containing the most primitive of living insects. The 850 known species have been called living fossils, for they are far older than the dinosaurs. Most thysanurans are elongated insects with normal biting mouth parts and three long spinelike processes at the tail-end of the body. The most primitive thysanuran is *Tricholepidion gertschii*, a secretive little creature that lives beneath bark and among debris in California.

These elusive and little-known insects are found virtually worldwide in damp, humid situations, among rocks, in caves, and in buildings. Only about four of the 47 silverfish species occurring in North America are moderate pests in homes and libraries, among other places, as they have a fondness for glues, papers, and starchy materials. The two so-called "domestic" silverfish species are *Lepisma saccharina*, and the "firebrat," *Thermobia domestica*. They differ in that the former prefers damp, cool locations while the latter prefers warm, dry places, such as restaurants, homes, and other centrally-heated locations.

Unless present in great numbers through poor sanitation practices, these insects seldom constitute a major health, sanitation, or even nuisance threat. The majority of the silverfish species in this group live outdoors, and are rarely encountered by people. Many live in symbiosis with termites and other subterranean, social insects.

Field cricket
Gryllus assimilis

The big-headed, glossy black field cricket is a familiar sight almost everywhere there is debris, weed growth, or a house to shelter it. Found virtually throughout North America, the field cricket's shrill, intermittent song is a familiar, late summer sound. This cricket is easily reared in captivity, feeding on lettuce, bread, bone meal, and water, and is commercially raised in great numbers as a research animal and as food for lizards, snakes, and turtles. Millions are sold as fish bait and as live food for pet reptiles and amphibians every year.

The native field cricket often invades houses and barns in order to hibernate, but the imported European house cricket (*Acheta domestica*) is the species more often seen in the cellar or hopping fitfully near the foundation. This cricket is similar to the field cricket but is paler brown and a bit smaller; it is a serious food pest whenever it enters homes, cheery song or not.

The field cricket, as well as the long-horned and tree crickets, have long been used as fairly accurate indicators of air temperature. The calls of so-called "temperature crickets" tend to slow down in cooler weather and speed up when it's warm, so that by counting the number of a cricket's chirps over a period of 15 seconds and adding the figure to the number 37, a good approximation of the current air temperature can be calculated.

Broad-winged katydid

The long-horned grasshopper, known also as the broad-winged katydid (*Microcentrum rhombifolium*), is common throughout the southern United States, occurring as far north as Utah, Kansas, and New York in fields, gardens, and waste areas. This attractive, bright green grasshopper feeds on a wide variety of trees and shrubs and is often a serious pest of ornamental shrubs.

The similar fork-tailed bush katydid (*Scudderia furcata*) is common throughout much of North America, frequenting a wide variety of habitats, from marshes and meadows to golf course edges and abandoned, overgrown landfills. Only the males sing; the females have special auditory organs on the bases of their front legs used to hear and home in on their amorous, summer-night suitors.

European earwig
Forficula auricularia

This essentially nocturnal scavenger was introduced into the United States in the early 1900s and has since spread from coast to coast. Despite the common name, of medieval origin, there has never been a verified report of the creature invading a human ear.

Earwigs are common virtually anywhere there is thick vegetation and natural or manmade rubble and debris. They can be serious garden pests—they will devour roses and other flowers as well as some fruits and vegetables. The abdominal forceps or pincers of a large male earwig might look dangerous,

Earwig

but the pinch is rather minor and ineffectual. Male pincers are curved, while those of the smaller female are straight. Although earwigs are winged insects and capable of flight, they seldom fly. They are reportedly attracted to lights, however, and are found buzzing clumsily about and bumping into porch lights and screens on warm summer nights.

Earwigs rank high among those insects that inspire feelings of disgust and fear among people who encounter them. Because these insects hide under all kinds of objects, and especially favor dark, damp nooks and crannies in cellars and garages, as well as the folds of tents, piled clothing, and camping gear, encounters can be frequent. The earwig's longish, shiny, sinuous form and deliberate, snaky manner of crawling about lend them a vaguely menacing air, small as they are. When touched or molested, an earwig quickly raises its abdomen and spreads its impressive pincers wide, ready to strike if bothered further.

Earwigs are rather unusual among insects because they engage in both brooding their eggs and short-term care of the young. The female deposits her eggs in a dark, secluded place and curls her body about the clutch in a protective posture. She similarly watches over the hatchlings for a few days after their emergence, until they are active enough to strike out on their own.

Mosquitoes
Anopheles and Aedes genera

48

There are more than 2500 species of mosquitoes worldwide, but only a few of them are serious pests of mankind. Many species of mosquitoes make general pests of themselves through their biting activities, but the real complaint is their function as vectors of potentially lethal diseases such as malaria, yellow fever, dengue, and others. They are difficult to control, the only effective methods unfortunately revolving around the use of chemicals of some toxicity. DDT is still used in the Third World to control them, much to the detriment of the environment, but doubtless saving many lives. The larvae of all of them are aquatic—often called "wrigglers"—and they feed on plant material such as algae, bacteria, fungi, and decaying vegetation. The so-called malarial mosquito (Anopheles quadrimaculatus) occurs throughout southern Canada and from the eastern U.S. to Texas. It is dark brown to nearly black, with four rather conspicuous white spots on the wings. The larvae prefer clean still waters with plenty of shade. This mosquito was the principal vector of malaria in this country up until World War II; malaria is transmitted only by species of mosquitoes of the genus Anopheles.

The infamous saltmarsh mosquito (Aedes taeniorhynchus) needs little introduction to those living near or vacationing at the seashore—it is prevalent near coastal areas throughout North America. The female dines on blood, the male on nectars and plant juices. This mosquito is often naturally controlled by brackish water fishes, such as the mummichog, sheepshead minnow, and the aptly named mosquito fish, Gambusia affinnis (see chapter 4, "The fishes").

The Asian tiger mosquito (Aedes albopictus) is a relatively recent arrival in North America, first reported in Houston, Texas in 1985. A mere eighth-inch in length, this attractively patterned, black and white little hummer is highly aggressive and very persistent—even for a mosquito—and serves as a vector for eastern equine encephalitis (EEE), an especially virulent and usually fatal brain affliction. A. albopictus has long been identified as a carrier of dengue, or breakbone fever, a painful but nonfatal form of encephalitis prevalent in the tropics. An outbreak of dengue in Hawaii early in this century was attributed to this mosquito. Until recently, however, it had not been thought to carry any diseases threatening to humans in the continental United States.

The mosquito is not host-specific, that is, it attacks a wide variety of warm-blooded animals that might in turn serve as transmitters of the deadly EEE virus to people. The tiger mosquito, which arrived in Hawaii around the turn of the century, has spread to 20 southern and midwestern states, and its presence is suspected as far north as Minnesota and New Jersey. According to scientists, it has definitely become established in Florida, as well as in Alabama, Arkansas, Delaware, Georgia, Illinois, Indiana, Kansas, Kentucky, Louisiana, Maryland, Missouri, Mississippi, North Carolina, Ohio, Oklahoma, South Carolina, Tennessee, Texas, and Virginia. Although an individual tiger mosquito is not a strong flyer and seldom wanders far from its birthplace, the prevailing southwesterly winds of the North American landmass have doubtless aided and abetted the insects' spread to the north and east.

In yet another example of people serving as unwitting agents in the transport of pest species, the tiger mosquito is thought to have arrived on this continent via water contained in old, used tires being shipped from Japan for the domestic recapped tire industry. In fact, this creature actually favors manmade breeding sites in the form of birdbaths, open buckets, empty cans and bottles, and anything else that will hold water, thus putting it in potential close contact with urban humanity. The worst outbreak of the EEE virus occurred in Polk County, Florida, and was centered in the area surrounding a huge tire dump not 12 miles from Disney World. The disease is fatal in nearly 80 percent of its human victims, and leaves the survivors in a disabled, brain-damaged state. Scientists feel the massive 1991 outbreak of the EEE virus among horses throughout much of the southeast including Florida might be traceable to this pest.

The tiger mosquito is native to temperate and tropical Asia and Japan and might well expand its range northward in North America if the current warming trend continues.

Fruit fly
Drosophila melanogaster

Fruit flies are everywhere, both environmentally and geographically, for they invariably turn up anywhere foods or fruit are stored, and are found all over the world in more temperate climes. For the most part, fruit flies are little more than a casual nuisance, though they can serve as vectors of certain fungus diseases of plants. Tiny insects, they can easily gain entry into a home through the mesh of ordinary window screening.

The fruit fly has long been used in classrooms and laboratories in the study of genetics. Multiple mutations of the original have been produced through selective breeding, and much of what we know of the principles of genetics has been gained through comprehensive studies of this tiny creature over the past hundred years.

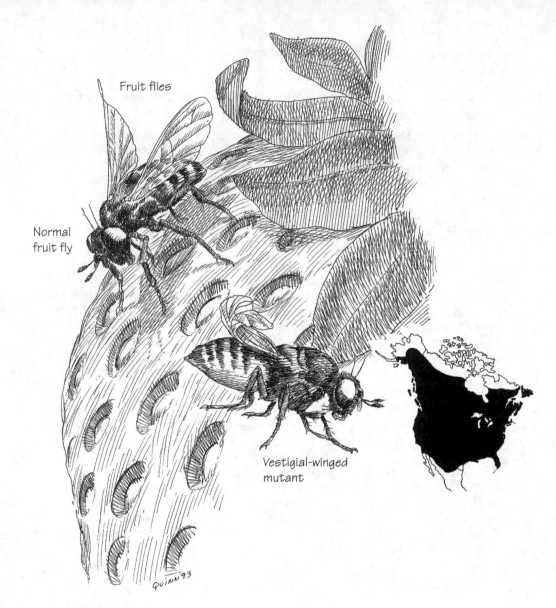

Fruit flies

Normal
fruit fly

Vestigial-winged
mutant

QUINN '93

Houseflies
Musca domestica

The common housefly needs no introduction anywhere in the world. This insect (and the various species of blow flies and flesh flies) are ancient and ubiquitous pests long regarded as serious threats to human well-being, contaminant agents, and the vectors of many diseases, in particular typhoid fever. They are found worldwide, everywhere that man calls home and prepares, stores, and discards foodstuffs.

The procreative prowess of the housefly is astounding. The flies breed in decaying food materials, offal, and excrement. The average female fly lays from 150 to 250 eggs, producing roughly a thousand ova during the insect's lifetime of five or six months. It has been estimated that a pound of meat left exposed on a warm summer day will accumulate as many as 165,000 fly eggs. The cycle

Housefly

from egg to larva to pupa to adult encompasses about two weeks. It has been speculated that a single pair of flies, instituting reproduction in April, would be the progenitors of some 191,010 trillion neophyte flies by August if all of their offspring were to survive the myriad threats arrayed against them. Few insects have more enemies lined up to make a meal of them than the fly, a factor that prevents us from finding ourselves literally knee-deep in flies at any given time of the year.

There are more than 85,000 described species of true flies (those flying insects possessing a single pair of wings, as opposed to two pairs) worldwide, with some 15,000 occurring in North America. Butterflies, mayflies, dragonflies, and fireflies are not true flies, while crane flies, flower flies, stable flies, robber flies, and houseflies are. The great majority of flies are at best beneficial animals and at worst harmless to mankind's interests. All serve a profound and vital purpose in the scheme of things, regardless of our perception of their activities. The flies run the gamut from the harmless and ethereal crane flies and midges that dance airily in the summer evening, to the pestiferous biting midges, often called "no-see-ums," black flies, and the formidable horse and deer flies of summer outings.

The blow flies (*Lucilia* species), also called bluebottles and greenbottles, are bright metallic green or blue, rather large flies that are prevalent around fresh and decayed meat, garbage, and excrement. They are attracted to open, poorly treated wounds, and the maggots will eat both necrotic tissue and living flesh.

The flesh fly (*Sarcophaga haemorrhoidalis*) occurs virtually all over the world in temperate and tropical climes and is an obnoxious pest of foods. The larvae are scavengers, living on all manner of decayed material; they breed in human excrement as well as in the human digestive tract.

One of the more intriguing "fly control" stories is China's recent, determined campaign to transform the capitol of Beijing into a "City of No Flies" (*The New York Times*, September 7, 1992). Government officials, hoping to interest the International Olympic Committee in the choice of the city as the site for the 2000 Olympic Games, decided that Beijing's chances would be much improved if the place were freed of the billions of houseflies that call it home. Beijing, like all major cities the world over, is a veritable housefly heaven but the authorities resolved that the insects could be successfully evicted through the application of more than 130 tons of pesticides and a dedicated populace wielding 200,000 government-issue fly swatters.

The anti-fly assault began in March 1992 and consisted of a series of heavily publicized fly-zapping orgies called "attack weeks." In one such campaign, 1000 teams of children and elderly people formed special "swat teams" and fanned out over the city swatting flies, spraying poison, and policing dumpsters and public areas prone to the accumulation of litter and garbage. More than 15 tons of pesticides were lobbed at the flies and some 270,000 tons of fly-producing garbage was disposed of. The ultimate "fly density standard" for the city is an optimistic two flies permitted in one room out of 100, probably an impossible level to maintain indefinitely given the housefly's astonishing capacity for reproducing itself, and the city's human, trash-producing population of more than 10 million.

The Chinese effort at fly control, despite huge odds against its complete success, is just about the only way to effectively eliminate the housefly from the urban scene today. An antiseptic environment is the bane of the housefly, and the heart of any anti-fly campaign can be found in the words of a senior Beijing health official: ". . . we won't just kill flies. We want to create clean cities."

Housefly eggs are laid in garbage and other decaying matter and hatch in one day. The larvae mature in five to seven days, pupate for another seven, and then emerge as adults. The adult flies feed for about two weeks before mating; 10 or more broods may be produced in a season, depending on the local climate.

Ants and termites

The industry, organization, and apparent intelligence of the ants has inspired mankind for many ages. Among the lower animals, the ants possess what has been called the social order closest to that of humankind's, the great, swarming "ant-hill cities" of the present day an apt illustration of that claim.

The Bible admonishes the indolent among us to "Go to the ant, thou sluggard; consider her ways and be wise: which, having no guide, overseer, or ruler, provideth her meat in the summer and gathereth her food in the harvest."

The word *ant* itself stems from the Old English *aenete*, which is akin to the High German word, *ameiza*, meaning "the cutter of." The active, highly social, wood-nesting carpenter ant (*Camponotus pensylvanicus*) is closely associated with man and his dwellings, and is one of the more familiar species in nearly all regions. Its color varies between light red and brown and the more commonly seen glossy black. The carpenter ant occurs throughout eastern North America and west to Texas. The nest is excavated out of the wood of dead or dying trees, stumps, and the frames and foundations of houses. A colony might contain as many as 2500 workers, some males, and a single queen.

Carpenter ants feed on dead and living prey and are attracted to fruits and their juices. They can infest a house virtually unnoticed and their extensive galleries can extend far into the adjoining soil.

The mound ant (*Formica pallide-fulva*), also called the red or common house ant, lives primarily in the soil, building mounds of dirt and bits of refuse that give away the colonies' locations. The species is distributed throughout the United States west to the Rocky Mountains. The pavement ant (*Acanthomyops tetramorium caespitum*) is another very common species found nearly throughout the temperate zones of the world. This small red ant is the species most often seen near cement house foundations and on sidewalks, where their extensive underground galleries can undermine the concrete and cause plenty of damage to buildings and walkways.

Carpenter ants

The ant species perhaps best known to the average person is the aggressive fire ant (*Solenopsis geminata*), of South American origin. Accidentally introduced into the United States via Mobile, Alabama in 1918, this species had spread to nine other southern states by 1957, and today is found throughout much of the southern half of the country. Due to its large nest mounds, aggressive nature, and ferocious bite, the fire ant became the target of a vast federally-funded eradication campaign commencing in the late 1950s, and costing an estimated $7 billion. The fire ant eradication program involved the heavy application of the dangerous pesticides dieldrin and heptachlor, which resulted in much environmental havoc and the destruction of beneficial organisms, including countless millions of songbirds. Critics felt that the supposed fire ant threat was much less than perceived, and the campaign was eventually ended.

Nearly all species of ants have a highly-developed social order and display an "intelligence" and sense of organization rare among the insects. The life cycle includes a colony consisting of a queen and her sterile workers overwintering deep within a hibernaculum. In the spring the queen resumes egg laying and the workers fan out to seek food for the entire colony. The larvae pupate for about three weeks in most species, and the winged males and females leave the colony to fly about and mate. The males die, but the fertilized females, future queens, retire underground and found new colonies. Ant colonies gradually increase over the years under favorable conditions; workers can live for up to six years, queens for as long as 15 years.

In the end, in the timeless words of Auguste Forel, physician and dedicated ant-watcher of the last century, "The ants' most dangerous enemies are other ants, just as man's most dangerous enemies are other men." These fascinating creatures seem destined to survive in the urbanized world of tomorrow as long as food and shelter remain available to them and their living space is free of the toxic agents that are, ultimately, hostile to all life.

Another group of insects often mistaken for ants but completely unrelated to them are the subterranean termites (*Reticulitermes* species). The ants are members of the order *Hymenoptera*, which includes the bees and wasps, while the secretive, though equally social termites are placed in the *Isoptera*.

Although they resemble each other superficially, termites can be readily distinguished from ants by their lack of an antlike waist constriction at midbody and their antennae, which are straight as opposed to angled, as in the ants. Ants have large, well-developed eyes, while those of termites are small or completely absent. There are two North American termite species that can cause damage to wooden structures: *R. hesperus* is found throughout western North America, and *R. flavipes* is found in the east. Both species are entirely subterranean and require ground contact from which to bore tunnels into wood; faulty construction practices are the cause of much of the termite infestations. Wood buried at a building site can serve as the nucleus of a colony, and supports should always be treated where they contact concrete or the ground.

Termites are elongated, pale-bodied creatures that function within a complex caste system. The colony's queen might be many times the size of her subjects, and the heads of the soldier termites are greatly enlarged and armed with formidable jaws. The workers are soft-bodied, usually lack functional eyes, and do not develop wings. The many termite species worldwide construct elaborate nests is soil, in trees, and in dead—never living—wood. Although they do great damage to wood by chewing it in the construction of their extensive galleries, termites are unable to digest its cellulose as food without the aid of symbiotic bacteria that live in their intestines. The bacteria provide the enzymes necessary for the conversion of indigestible cellulose molecules into usable carbohydrates.

An extinct termite, Mastrodermes electrodominicus, yielded the oldest DNA ever recovered from a fossil. Scientists extracted the fairly long sequences of nucleic acids, the building blocks of genes and genetic inheritance, from a termite and a stingless bee, both found embedded in amber carbon-dated back 30 to 40 million years. The specimens were collected in the Dominican Republic, in amber formed from the sap of an extinct tropical legume called Hymenaea.

Japanese beetle
Popillia japonica

This very destructive beetle was introduced into the United States from Asia in 1913, and has since spread slowly across much of the eastern United States. The first officially reported place of infestation was in Burlington County, New Jersey, where inspectors of the state's department of Agriculture collected about a dozen beetles in 1916. Three years later it had increased its numbers and spread so extensively throughout much of the state that one researcher collected 20,000 of them by hand in a single day. It is thought that the Japanese beetle arrived in North America in the grub stage, via the soil surrounding the roots of ornamental plants imported from Japan and other Asian countries. The Japanese beetle is a member of the Family *Scarabaeidae*, sharing it with the famous scarab beetles of Egypt and North Africa that figure prominently in ancient religious myths and customs.

Japanese beetle

The eggs of this beetle are deposited in the soil, and the larvae feed on the roots of grasses; after a two-year larval stage, the grubs pupate into adults in midsummer and the beetles then infest a wide variety of plants, in particular those of the rose family. These beetles often damage fruit trees, defoliating an entire tree if the infestation is severe.

The Japanese beetle has been controlled by contact-poison insecticides and by use of the once-familiar hanging "Japanese beetle traps" still in regular use in agricultural areas in much of the South. It is controlled naturally by a fungus known as the milky disease, and by several species of parasitic wasps and flies. In addition, the starling, another exotic import and avian survivor par excellence, is an important predator of the beetle, and figured prominently in its eventual control.

Cabbage butterfly
Pieris rapae

The imported cabbage butterfly is without a doubt the butterfly species most familiar to Americans and Canadians. The great majority of people who see it flitting erratically but determinedly over

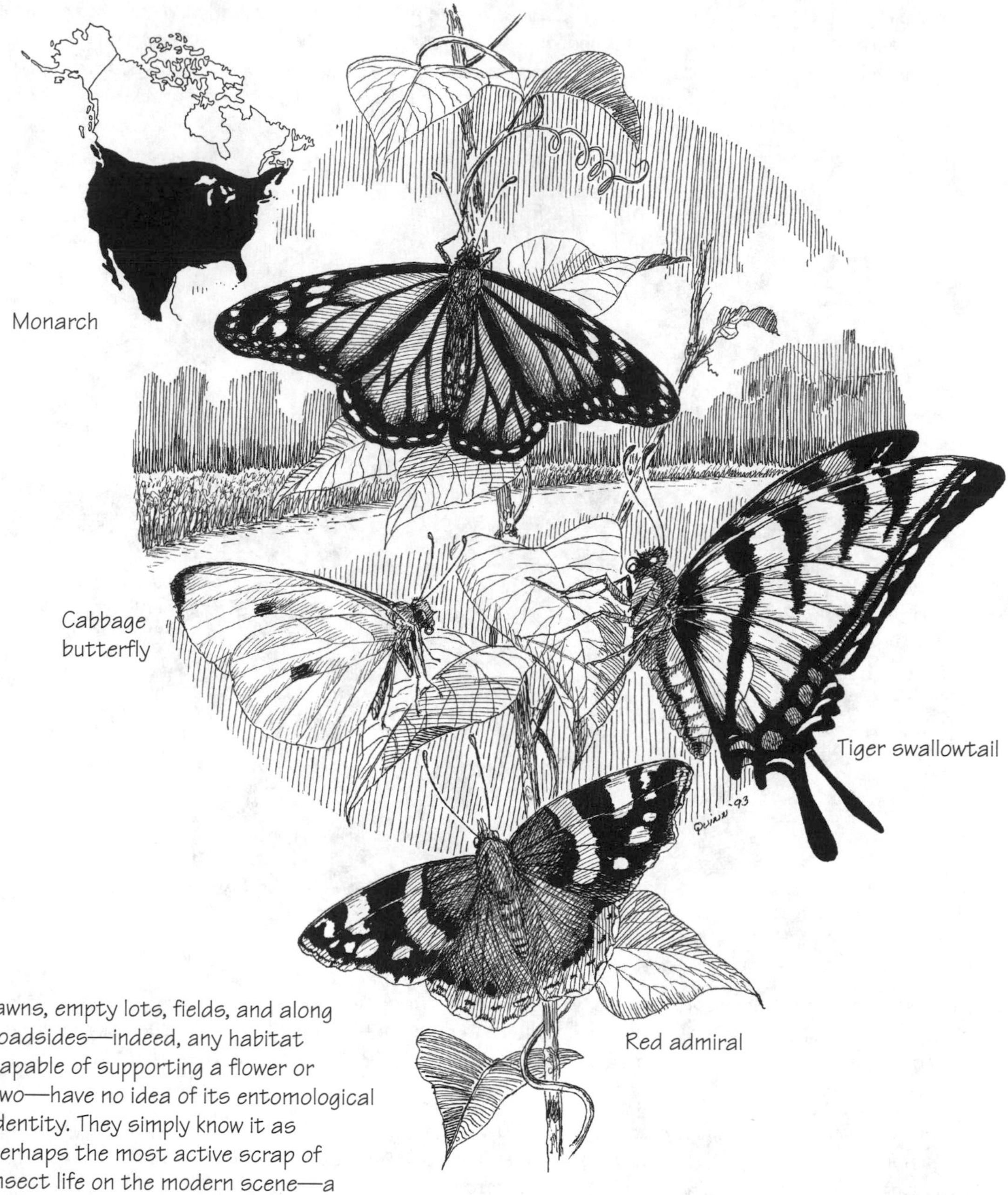

Monarch

Cabbage
butterfly

Tiger swallowtail

Quinn-93

Red admiral

lawns, empty lots, fields, and along
roadsides—indeed, any habitat
capable of supporting a flower or
two—have no idea of its entomological
identity. They simply know it as
perhaps the most active scrap of
insect life on the modern scene—a
small white butterfly that seems to be everywhere.

The cabbage butterfly, also called the cabbage white, was introduced into Quebec by accident in
1860, and eight years later it turned up in New York. By 1900 the insect had spread throughout the
eastern half of the continent and today is found from coast to coast, from southern Canada to
Mexico in all but the driest of habitats. The small green caterpillars feed on a wide variety of plants
but have a definite preference for cabbage, hence the common name. This is one of the few butterfly
species that are serious pests on garden produce.

Cabbage butterflies are highly conspicuous insects wherever they are found. They frequently engage in spirited upward spiral flights involving two butterflies that might rise as high as 60 feet in the air. These flights are usually the evasive tactics of an already mated female trying to shake a courting male so she can get back to the business of egg-laying. The small green larvae are often seen prowling over the leaves of cabbage, mustard, broccoli, and other garden plants.

There are two other species in this genus, both called whites and both native to North America. The checkered white (*P. protodice*) is found throughout the southern states, though its numbers have declined greatly over the past 50 years in the face of competition with the feisty European immigrant. The old-fashioned, or mustard white (*P. napi*) occurs in the northern states but moved southward following the arrival of the cabbage butterfly. In an apparent attempt to avoid competition with the aggressive interloper, the mustard white abandoned its former open fields and forest edge habitats and literally "took to the woods." It is now primarily a woodland butterfly wherever it is found.

The closely related sulphurs (*Colias* species) are widespread throughout North America, persisting in the face of urbanization as long as there are abundant clover and flowering weed crops to be found in fields and vacant lots. The common sulphur (*C. philodice*) occurs throughout North America except in the Arctic, the deserts, and in tropical Florida. The alfalfa butterfly (*C. eurytheme*) is larger than the preceding species, and orange rather than yellow in color. It occurs throughout the continent, though it does not extend as far north as the common sulphur. The dog face butterfly (*C. cesonia*) occurs throughout most of the United States and south to Argentina. The larvae feed on legumes, such as red and white clover and soybeans, but their numbers and combined appetite seldom add up to an insect-pest problem.

Other adaptable butterflies

The tiger swallowtail (*Papilio glaucus*) is abundant virtually everywhere throughout eastern North America, found in extensively wooded and agricultural regions as well as in city parks. The larvae feed on a wide variety of plants, including ash, basswood, birch, poplar, cherry, maple, apple, and others. The western tiger swallowtail (*P. rutulus*) is found throughout western North America eastward to the Rockies. Its larvae feed on willow, alder, sycamore, and aspen.

The red admiral butterfly (*Vanessa atalanta*) is an active, strikingly-colored butterfly occurring throughout North America, and indeed, widespread throughout the northern hemisphere north of the equator. The larvae feed on nettle and hops.

The rather somber-hued, though lovely mourning cloak butterfly (*Vanessa antiopa*) is found throughout the temperate regions of the northern hemisphere, occurring in the western hemisphere as far south as Guatemala. Known as the Camberwell Beauty in England, the mourning cloak has, for some reason, declined in the British Isles over the past century and today is rarely seen there. As is often the case with rare and endangered creatures, their very scarcity makes them all the more appealing to collectors, who hunt down or purchase the last individual of a species in order to add it to a lifeless collection. That this is so in the case of the mourning cloak is illustrated by a reportedly true incident early in this century in which a British butterfly collector imported several mourning cloak larvae from the United States, reared them to maturity in captivity, and then released and immediately recaptured the adults in an English woodlot. All so he could claim that he had officially collected the rare species on English soil!

This butterfly is one of the few species that hibernates as an adult, emerging very early in the spring to flit through the leafless glades of still wintry woodlands. The larvae feed on willows, poplars, and elms.

Second only to the cabbage butterfly as the most familiar butterfly in North America, the monarch, or milkweed butterfly (*Danus plexippus*) shares the number two slot with the big, equally flashy tiger swallowtail. The monarch's size, bright orange-and-black color pattern, and its habit of flying boldly and yet leisurely over open areas, across busy highways, through urban centers, and even across sea beaches combine to make it one of the most noticeable insects today. Occurring widely from southern Canada south throughout the United States, this butterfly undertakes arduous migrations to coastal Southern California and Mexico, where it winters in several isolated valleys protected from human alteration and destruction. The larvae feed almost exclusively on the several species of milkweeds, leading to the animal's other common name "milkweed butterfly." It is poisonous in both the larval and adult forms and thus scrupulously avoided by predators.

Mourning cloak

Common sulfur

Common sulfur

The monarch butterfly has been able to adapt to—and even thrive in—the considerable alteration and modification of the breeding and summering habitats that have taken place throughout temperate North America. However, the species will be hard pressed to persevere in the face of the destruction of its isolated and concentrated wintering grounds. Illegal logging in the Mexican coastal state of Michoacan, carried out by clandestine sawmills in defiance of the 1986 presidential ban on the practice, has, according to some reports, resulted in the deaths of some 39 million monarchs, nearly 70 percent of the estimated total 1991–92 population of the insect. Mexican police have closed 15 of the mills and arrested their operators, but doubtless such illicit logging will go on unless the wintering sites are placed under round-the-clock surveillance and strictly guarded, much as some nations have been forced to protect vulnerable sea turtle breeding beaches from human exploitation. In the United States, developers continually apply pressure to open up coastal California monarch wintering areas to development; if these efforts are eventually successful, the outlook for the monarch butterfly, still a common sight in American suburbia today, will suddenly become very grim indeed.

Tent caterpillar
Malacosoma species

Tent caterpillars exist in about 35 species throughout North America. The adult is a small moth, recognizable by its rich yellow-brown color with two dark, narrow bands across the forewings. They are important pests on both forest and cultivated trees and shrubs, preferring apple, wild cherry, and others

By far the most abundant and familiar of the tent caterpillars is the eastern tent caterpillar (M. americana), best known for its often extensive white web colonies festooning forest and suburban shade trees. The adults of this species are pale brownish, stubby-bodied little moths with a wingspan of about one and a half inches; at this stage, the insect is virtually unknown to most people, but the larvae certainly are not. The females lay their whitish egg clusters on the twigs of their food trees; the caterpillars emerge in the spring when the trees are beginning to leaf. Soon after they begin feeding, all of the caterpillars that emerged from an egg case assist in building the web nest. This is usually located in the crotch of two or more branches of the host tree and is composed of tough layers of silk with air spaces between each layer.

Although to the casual observer tent caterpillars seem to wander randomly to and from the web, they have, in fact, a definite purpose and order to their activities. First, the nest is built in stages, with new layers of webbing added as the caterpillars grow in size. The larvae usually leave the nest to feed three times a day: in the early morning, at midday, and just after dusk. Each time they depart on feeding forays over the host tree, the caterpillars leave a trail of silk on the branches as a means of marking the route back to the nest. These trails are thicker on the "main highways" of the larger branches and slimmer on the twigs, the "secondary roads" on the tree. A tent caterpillar stops eating only on cooler days, or when it undergoes one of the five molts it goes through in its larval life. When it's time to pupate into the adult moth, the caterpillars abandon the nest and wander about in search of a protected spot, usually beneath loose bark or in a natural or building crevice; many cross roads at this time and are flattened by passing cars.

Although a number of bird species will eat tent caterpillars, the decline of the black-billed and yellow-billed cuckoos in North America has made life easier for this insect pest. One predator it still has to contend with, however, is a small tachnid fly that darts onto the web nest and deftly lays an egg just behind the head of a handy caterpillar, a spot where the larva cannot remove it easily. The emerging fly larva burrows into the body of the caterpillar and feeds on its tissues and fluids, eventually killing it. Tent caterpillars have been observed energetically flicking their heads back and forth in an effort to thwart the egg-laying attempts of this determined little fly.

Tent caterpillar

Gypsy moth

Gypsy moth
Porthetria dispar

This tree-devouring entomological menace occupies a prominent position in the popular consciousness, if only because of the volume of media hype surrounding its rapacious depredations and the often desperate human campaigns to combat them. All of the many billions of gypsy moths that today give foresters and pest control agencies in 17 states their yearly migraines can be traced back to a few individuals that escaped from a private collection in 1869. The story has it that French entomologist Etienne Trouvelot, conducting research in eastern Massachusetts, imported a supply of eggs and caterpillars from Europe in order to culture the species as a possible alternate source of silk. His thought was that cultivated strains of the gypsy moth—a known pest in Europe—might prove to be resistant to silkworm diseases sweeping Europe at that time and playing havoc with the silk industry there. Alas, security was something less than tight in his enclosures and, according to entomologist Frank Lutz, perhaps understating the event in 1918, "some of the specimens went off and started to colonize America."

A contemporary account of the first reported outbreak of the moth in nearby Medford, Massachusetts 20 years after the escape graphically documents this insect's reproductive and gustatory abilities:

> The numbers were so enormous that the trees were completely stripped of their leaves, the crawling caterpillars covered the sidewalks, the trunks of the shade trees, the fences and the sides of houses, entering the houses and getting into the food and into the beds . . . . The numbers were so great that in the still summer nights the sound of their feeding could be plainly heard, while the patter of their excremental pellets on the ground sounded like rain. Valuable fruit and shade trees were killed in numbers by their work, and the value of real estate was very considerably reduced. . . .

Among other trees, the gypsy moth favors several species of oaks as food, and upon escaping, they quickly spread southwestward across New England and into more agreeable climes. The first major moth outbreak in nearby Medford Massachusetts in 1889 was just the beginning of a long, ongoing entomological nightmare. Today, despite an intensive spraying campaign, the gypsy moth is firmly established as far west as Michigan, throughout much of Ohio and West Virginia, and as far south as parts of North Carolina. The infestation has spread, on average, at the rate of between 15 and 30 miles a year, aided primarily by winds, but also hitch-hiking on motor vehicles using the nation's helpful network of roads and interstate highways. More than 7.5 million acres nationwide were defoliated by the determined and voracious little creature in 1990 alone.

No one knows for sure just how much the control campaign against the gypsy moth has cost over the years—many billions of dollars, no doubt—but as of 1990, the Forest Service, which primarily combats the larger infestations as they occur, spent an average of $6.1 million a year since 1979—between $7 and $20 an acre—on combatting the insect. In 1990, the Service shelled out $13.6 million to keep the pest at least partly at bay.

In the early 1980s the chemical pesticide Sevin was used to combat the gypsy moth, but concerns about its effect on the overall ecosystem has prompted a switch in most states to a kind of bacterial warfare against the moth. *Bacilllus thuringiensis*, or *Bt*, inflicts a massive case of indigestion on the larvae when they attack vegetation treated with the chemical. Though it is not harmful to most other insects, wildlife, or livestock, the compound is fatal to the larvae of nearly all other harmless butterflies and moths, and the massive use of the stuff just might be one of the factors behind the gradual disappearance of butterflies from the summer scene throughout most of the East.

The gypsy moth larvae devour an astounding 500 species of plants, including conifers. In New Jersey, the gypsy moth defoliated 431,000 acres of woodland in 1991, up from 137,000 acres in 1989. The 1988 infestation of 74,000 acres represented a low point in the population cycle, which had been declining steadily from the 1981 peak of 800,000 acres affected. The insect is favored by dry spring weather, and wet years usually see a significant decline in the populations due to fungus infestations that control the moths naturally.

New Jersey's agricultural department has cultured parasitic wasps and other insects that prey naturally upon the moth. The natural predators effectively control moth numbers so that the cyclic peaks are now reached every eight years rather than the four years reached even under massive applications of pesticides. Aerial spraying costs about $13 an acre, with localities sharing the cost with the state and the Forest Service.

The private homeowner can protect individual trees without resorting to toxic insecticides by:

- using Bt sprays available from local garden supply centers.
- girding trees with insect-catching, or "tanglefoot" compounds that work like old-fashioned flypaper to nab the caterpillars.
- wrapping the trees with a burlap "skirt" just below the lowest branches. During the day caterpillars will retreat down the trunk and take shelter under the burlap, where they can be ambushed and scraped off the tree and into a bucket of soapy water to destroy them.

Praying mantids
Stagomantis species

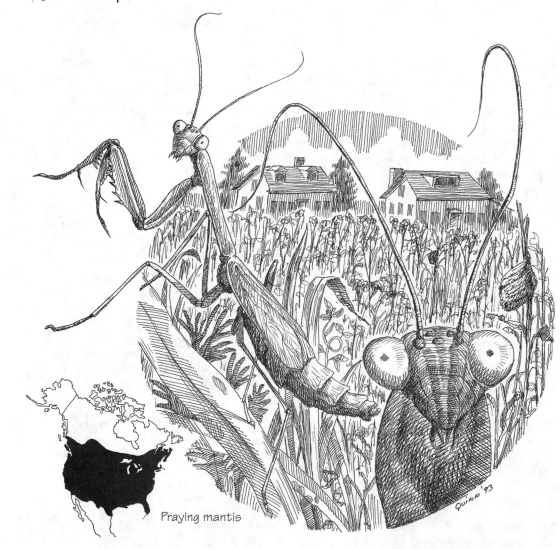

Praying mantis

The familiar praying mantis is unique on at least two counts: it is one of the few known insects in the world that can turn its head and look over its own shoulder; and it is the subject of one of the most enduring folklore myths in the United States. The myth is that the mantis is a federally protected animal and anyone bothering or squashing one is subject to a $50 fine if caught in the act. While the first item is true, there is no law anywhere on the books that protects the mantis from human swatters and stompers. When and how that tale got started is a mystery even to folklorists. I can clearly recall the phantom regulation from childhood in northern New Jersey in the 1940s; the reported fine was always and specifically $50, no more and no less, in all adult warnings against molesting the helpful insect.

Some entomologists speculate that since the praying mantis is an entirely beneficial, voracious creature that destroys hordes of insect pests (along with beneficial bugs and butterflies whenever it can catch them) and is a large, spectacular, basically attractive animal that garners interest whenever spotted, the desire to shield it from harm developed almost inevitably and naturally. The insect's head-turning ability and the "praying" attitude assumed by the large, powerful, angular forelegs lends it the look of an animal rather than a mere bug in the eyes of many people, who would otherwise kill insects other than butterflies without a thought. The mantis is one of the few really intriguing insects in the eyes of the lay person, and its appearance and behavior invite feelings of cautious respect and protection.

The praying mantis is a member of the *Mantodea*, a division of the order *Dictyoptera*, which includes the grasshoppers and, perhaps ironically, the cockroaches. The animal is characterized by its large triangular head and the enlarged front legs, modified for the grasping of prey. Most of the 1500 species of mantids found worldwide are tropical in origin; several of the species occurring in North America are imports from Europe and Asia. The Carolina mantis (*Stagomantis carolina*) is the only native species found east of the Mississippi, while the European mantis was accidentally introduced via plant nursery stock in New York State in 1899. Another large species, the Chinese mantis arrived through intentional introductions made in the Philadelphia area about 1896. The two latter species are the larger insects, and by far the most noticeable whenever they turn up in gardens and along suburban streets. A fourth species, the California mantis (*Stagomantis californica*), occurs from Texas and Colorado west to California.

Although today the insects are almost universally known as praying mantis, in the past they were picturesquely called "devil's rear horses," "soothsayers," and, in the South, "mulekillers," due to the mistaken belief that the brownish fluid exuded by the mantis was toxic to livestock.

The characteristic globular or marshmallow-shaped egg mass is conspicuous among the dried weeds of late fall and winter, and can be collected to overwinter in a cool place. The nymphs hatch en masse, and must either be confined in a roomy space or immediately released in the garden, for they are strongly cannibalistic and will quickly catch and devour each other. They can be fed fruit fly larvae when very small, then adult fruit flies, and finally houseflies and other insects. If kept several to a cage, they must be fed very frequently or cannibalism will thin the ranks in short order!

The New Jersey Division of Fish, Game and Wildlife says it annually fields a surprisingly large number of calls from people reporting swarms of appealing baby mantids on their Christmas trees or scampering over the living room furniture. The tiny insects emerge in the warmth of a home from their egg masses, laid the past summer and early fall by the female on Christmas tree farm stock.

Paper and mud-dauber wasps
Polistes species

The various paper and mud-dauber wasps are among the more familiar flying insects of summer, living as they do in close proximity to humankind and its buildings. These wasps are members of the order

Hymenoptera, one of the largest and most diverse of insect groups, with more than 103,000 described species. More than 17,000 species of wasps, bees, and ants are known in the United States and Canada alone.

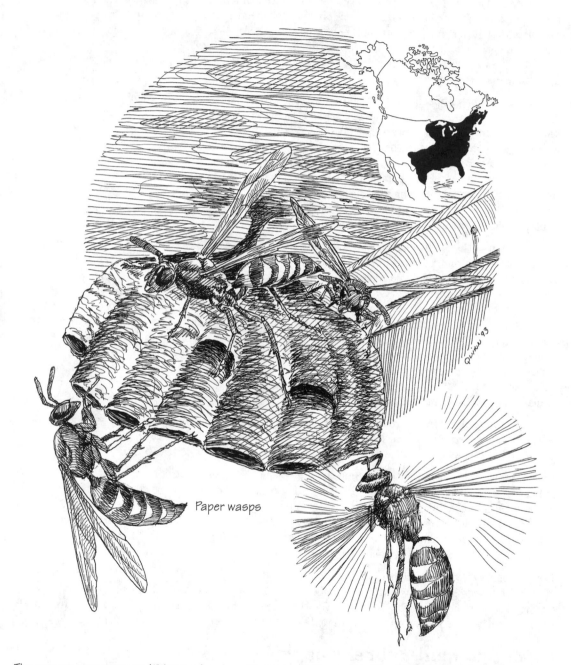

Paper wasps

The common paper wasp (*P. hunteri*) occurs throughout the eastern and southern United States as far west as Arizona. The nests, with the characteristic five-sided cells of the rearing combs, are familiar objects on the eaves and overhangs of porches and other outdoor structures. The nests lack the outer paper covering seen in hornet nests, and the comb is attached to the anchoring surface by a slim but tough *pedicel*.

The black and yellow mud dauber wasp (*Sceliphron caementarium*) is an abundant and ubiquitous species found through North America. It is closely associated with human habitations and has steadily expanded its range as urbanization has spread. The nest consists of several contiguous tubes of mud, each cell provisioned with spiders the wasps have captured. These wasps are often seen at pond edges or around roadside mud puddles, collecting the mud required for nest construction. They can be aggressive in the defense of their nests and can deliver a painful and repeated sting if molested. The ferocity of wasps in general was amply demonstrated in a report filed by a Vietnam War correspondent that the Vietcong had placed the active nests of tropical wasps across a road, effectively blocking the advance of government troops for a day.

One of the more common western mud-dauber wasps is *Pseudomasarus vespoides*, found throughout the western states and east to New Mexico, Colorado, and South Dakota. Many related species are common in the East. The mud nest is usually placed on a stout twig or on a flat rock surface.

Although the spider is the animal protagonist in the "Little Miss Muffet" nursery rhyme, it was actually a wasp that caused the little girl, a real-life person, to abandon her snack of curds and whey. The rhyme, penned in the late 17th century by one Dr. T. Muffet, was inspired by an incident that reportedly took place on a picnic in Epping Forest in England, during which Muffet's family was attacked by wasps. Muffet's daughter, Patience, was forced to drop her food and flee under the insects' determined assault, though the good doctor, unable to find a verse to rhyme with wasp, settled for spider instead. Muffet, an ardent amateur entomologist, contributed to *The History of Four-footed Beasts and Serpents*, edited by the Reverend E. Topsell and published in 1658.

Hornets and yellow jackets
Vespa and *Dolichovespula* species

The so-called aerial hornet (*D. arenaria*) builds large, familiar nests of masticated wood pulp (paper) in trees and low shrubs, under eaves, or in old buildings. In Europe, hornets are called wasps, a term applied to the paper wasps and mud-daubers in North America. This species is found throughout the United States and Canada as far north as the Arctic Circle. Both yellow jackets (*V. crabro*) and hornets are notoriously aggressive, and will spare no effort in the defense of the nest. People sensitive to the stings of these active and agile insects have been killed by concentrated attacks. The names hornet and yellow jacket are applied rather randomly to many species of black and yellow vespid wasps, but in general, hornets build large and often conspicuous paper nests on buildings or in trees, while yellow jackets set up housekeeping below ground, coming and going through carefully hidden entrance holes.

Hornets and yellow jackets overwinter as fertilized queens, starting a small, new colony in the spring. The queen constructs a few cells, lays an egg in each, and feeds the new larvae bits of captured prey. After a pupation period of about 12 days, the emerging adult, sterile, female wasps begin collecting food, enlarging the nest and caring for the new young. By late summer, the queen begins producing eggs that hatch into males and fertile females; these mate and the fertilized females will overwinter, beginning the timeless hornet cycle all over again.

Hornets and yellow jackets are much more active and visible in late summer and early fall as the social order of the colony breaks down and the wasps go their separate ways, feeding on fruits and other sweetish material. At this time they are frequent and uninvited guests at picnics and are often seen congregating around dumpsters, trash barrels, and on fallen apples and other fruits. During the early part of the summer the hornets are hunters of insects, which supply the required protein for the developing young. They are often seen flying about on the sunny sides of barns and other buildings looking for houseflies and other insects, and in the excitement of the chase they might zoom in at people, though seldom with evil intent!

Although generally labelled as noxious insects, hornets and yellow jackets generally mind their own business unless their homes are molested—not an unreasonable reaction to expect in any animal. A large, active nest might be located high in a tree on a busy suburban street, completely unnoticed through the warmer months by the human residents of the area, and spotted only when it is revealed in the fall.

Hornets

Bumblebee
Bombus americanus

Bumblebees of a number of species are common throughout the temperate regions of the world and are familiar to almost everyone, rural, suburban or urban resident alike. They occur in all manner of habitats, from urban lots to wilderness clearings, and have been observed in high mountain areas above the Arctic Circle. In North America *Bombus americanus* is the common bumblebee throughout the East, whereas *B. occidentalis* replaces it in the Far West. Bumblebees are closely related to honeybees, but unlike them, nest in the ground in the manner of yellow jackets. They are large, usually black and yellow or white bees with thickly-haired abdomens.

Bumblebees

Bumblebees set up colonies like most other bees and wasps, but they are usually smaller in the number of individuals and located underground. In the early spring the females buzz noisily about, foraging among spring flowers and searching for a suitable nest site. Such sites include the abandoned burrow of a chipmunk, mole, shrew or other small tunneling mammal. In more northern areas the competition for nest sites can be intense as the ground might still be partially snow-covered when the queens emerge. Studies have shown that in some areas fully 10 percent of all nests are taken over by another queen who either kills the rightful owner or forces her into submission. Bumblebees, which can sting but seldom do so unless handled, are extremely important pollinators of flowers and food crops. They are common and highly visible insects on lawns or in pastures where there is an abundant growth of red and white clovers.

Honeybee
Apis mellifera

The entirely beneficial and thus beloved honeybee is one of the most widely distributed insects, being cosmopolitan in the temperate regions of the earth. The species was brought to North America by the early European settlers, who saw it as a source of coveted honey in a strange, new land where there were no native honeybees. This immensely valuable insect is a member of the order

Hymenoptera, which includes such diverse insects as bees, wasps, and ants. Honeybees are today almost completely domesticated, and live in hives provided for them by man. They have, however, been known to leave these hives and set up wild colonies in hollow trees.

Honeybee

The life cycle of the honeybee is very well known. The insects overwinter in the hive as adults; the hibernating queen, workers (sterile females), and drones (sterile males) awaken in the first warm days of spring. The queen begins to lay eggs, which are tended by the workers. These hatch in about three days, and the larvae, fed on the honey produced from flower pollen gathered by the workers, mature in about six days. The larva's cell is then capped over in wax by the worker bees, and pupation into the adult bee takes place over 12 to 14 days. In summer the workers live about six weeks and the drones about eight weeks. Honeybee activity is particularly intense in late summer, when the insects swarm around goldenrod, asters, and other late-blooming composite flowers to stock up on food reserves for the coming winter.

The function of the honeybee in the ecological scheme of things is an almost perfect illustration of the often unnoticed, yet vital complexity of the goings-on in the healthy ecosystem. Most flowering plants must be cross-pollinated, either by broadcasting their pollen far and wide in a random manner, or directly, by insects or other animals that transport the male pollen from the stamens of one flower to the pistil, or female part, of another flower. This system of reproduction among plants ensures that a flower species maintains and enhances the genetic vigor required to adapt to and survive in a changing world.

Honeybees do not visit flowers in order to consciously aid in their reproduction; they simply seek to gather the greatest amount of food (pollen) with the least amount of effort. Therefore a given flower must develop strategies to attract bees in a field full of colorful rivals. Honeybees tend to make repeat visits to the blooms of a particular flower species once it has attracted them. This strategy on the bee's part, called *flower constancy*, minimizes the time consumed in delivering pollen to the hive, because it has learned where the pollen and nectar are located in a given flower. Observers have found that during the height of the summer season worker bees gathering food for the hive conserve time and energy by repeatedly returning to the same patch of flowers and even the same plant, and doing little or no casual exploring en route. The bees have developed other strategies to avoid visiting the same blossom twice, usually consisting of alighting near the base of the plant or at the edge of a large composite flower and working its way up or in a spiral motion.

Chapter four

The fishes

Pity the poor shad; who hears the fishes when they cry?
Henry David Thoreau

Among the vertebrate creation of North America, the fishes today face perhaps the greatest peril. Fishes and other aquatic organisms spin out their lives below the water's surface; any given species—especially the smaller, unobtrusive, nongame fish species—that pass quietly into extinction would be neither missed nor mourned by the average American today. Nearly 40 species, most of them endemic to the arid desert states and the Far West, have vanished over the past 100 years, 19 since 1964 (nearly one per year). Nearly 370 others—fully one third of some 1000 known North American fishes—are presently classified as Endangered, Threatened, or Of Special Concern by the Department of the Interior.

The statistics are discouraging and pretty much speak for themselves. Twenty-two of the imperiled fishes are endemic to Canada, 125 occur in Mexico, and the rest inhabit the continental United States. Many stocks of Pacific salmon are considered to be in dire straits due to degradation of habitats and dams, and acid rain has destroyed all of the known stocks of the aurora trout and severely degraded more than half the habitat of the Acadian whitefish. These species, of course, are game and food fishes and thus receive more attention from fisheries biologists, but many smaller, nongame species hover at the brink as well. The tiny Moapa dace has been reduced to less than 1000 individuals in a few pools and riffles in a single Nevada stream, while the surviving population of the celebrated Devil's Hole pupfish can almost be numbered on the fingers of two hands. The pupfish hangs on by a thread in a section of the Devil's Hole spring that is but 20 square meters in extent—the smallest known range of any vertebrate animal.

Since 1850, 67 percent of the fish species of the Illinois River have either declined drastically or been extirpated. Forty-three percent of the fishes of the upper Colorado River basin are extinct, endangered, or threatened. The fishes of the arid Southwest have been especially hard-hit; New Mexico lists 22 species as endangered, Arizona 22, California 42, and Nevada an astounding 43 species—the majority of it's native freshwater fishes. The Southeast, long recognized as having the greatest (62 percent) diversity of fish fauna on the continent, has suffered great losses due to increasing siltation of streams and acid precipitation. Tennessee reports 40 species in serious decline, Alabama 30, and Georgia 20; the equally grim situation in other states brings the total to at least 15 percent of the southeastern regional fish fauna classed as being "at risk."

The reasons for this alarming and accelerating decline continent-wide are many, but nearly all can be traced to human activities and the attendant population growth. Habitat degradation and alteration through increasing urbanization, chemical and agricultural pollution of aquatic systems, introduction of exotic species, and water diversion schemes have played havoc with fish populations; the specter of acid precipitation has placed at risk many lakes and ponds throughout the Appalachian Mountains and the entire northeastern United States and eastern Canada.

To all of the human-caused perils can be added the looming threat of a new natural enemy—a recently discovered dinoflagellate whose vast, red-tide-like blooms have been causing massive fish kills over widening areas of eastern coastal waters. Although dinoflagellates in general have gotten bad press as toxic agents, only a scant 42 of the some 2000 marine "dino" species are known to pose any health or environmental hazards. But within the past two decades, immense "blooms" of algae and dinoflagellates have wreaked havoc with the aquatic ecosystems of the Adriatic Sea and coastal Guatemala.

The toxic dinoflagellate was isolated and identified in 1988 after it invaded a laboratory colony of 300 tilapia and killed every fish. This microscopic creature, which multiplies in vast numbers in phosphorous and nitrogen-rich waters, has been tentatively given the species name *piscimorte*, or "fish killer." In only about 3 percent of those freshwater species currently considered at risk is commercial and recreational fishing considered a significant, human-caused factor.

Though it was identified and defined as a major environmental threat as recently as 1978, the grave threat posed by acid precipitation had its genesis 200 years ago with the birth of the Industrial Revolution, and has accelerated at a disturbing rate since about midcentury. Acid levels in rainfall have been rising rapidly in the southwestern states and throughout the Rocky Mountain region since 1950, and precipitation in northern Florida has become increasingly acidic over the past 25 years. Rain samples collected in several western states, including the supposedly pristine Colorado and Wyoming, have yielded readings ten times more acidic than unpolluted rainwater. In the province of Ontario alone, at least 50,000 lakes are thought to be at risk due to acid precipitation originating primarily in the United States. In Minnesota, one lake in six has been found to be dangerously acidic, to the point that noticeable declines have occurred among fish and invertebrate life.

Canadians are not blameless in acid production; the vast nickel and copper smelting complex surrounding Sudbury, Ontario, annually pumps some 4½ million tons of sulphur dioxide into the atmosphere. The huge plant is widely fingered as the single largest source of acid rain in the world, the nickel smelter alone emitting more sulphates in the 1970s than all of the volcanoes in history. Day in and day out, industrial mankind produces more and more sulphuric acid waste and other airborne toxins that drift across our skies like a malevolent, silent enemy, delivered to earth in the guise of that most innocent and welcomed phenomenon—the elemental gift of rain.

The fact that the fishes and other aquatic animals for the most part possess little of the charisma of larger furred and feathered vertebrates has inevitably made their protection an issue of somewhat lower priority. While people will quickly rally to the cause of the bald eagle, timber wolf, and blackfooted ferret, few fishes, with the exception of the notorious snail darter, have ever received much attention from mankind. But water is the stuff of life for us all, whether people or pickerel, and what happens to the fishes eventually happens to us.

Aquatic organisms, confined as they are to water bodies and thus more easily studied, are presently disappearing from North American habitats at a much faster rate than land-based fauna. The decline includes not only the more visible fishes, but such humble creatures as mollusks, crayfish, aquatic insects, and numerous other inconspicuous but nonetheless vital members of the food web. The Nature Conservancy estimates that fully one-third of the continent's fishes, two-thirds of its crayfish species, and nearly three-quarters of its mussel species are now rare or imperiled. Ten percent of North America's freshwater mussel species have become extinct since 1900, and 43 percent of the remaining species are classed as rare or imperiled.

Since the turn of the century, the greatest losses in fish species have occurred in the Great lakes drainages, the Great Basin, and the Rio Grande. Nearly all of these disappearances were due to the increasing demands of growing human populations on the scarce water supply. Today more than half of the freshwater fishes of the United States and Canada are protected throughout part or all of their ranges, a move that has slowed but certainly not halted the gradual slide toward extinction for many of them. The American Fisheries Society reports that between the years 1979 and 1989, 10 species of freshwater fishes are believed to have gone extinct. In that same decade, an additional 139 species have become endangered, threatened, or classed as "of special concern" due to loss of habitat and a subsequent sharp decline in population.

Despite aquatic conservation and species protection, it is likely that most native North American fish species will continue to decline, and many species and populations will probably vanish in the early years of the next century. The principal reason for this pessimistic forecast is that environmental preservation goals, no matter how well-meaning or extensive, are doomed to failure unless human numbers can be stabilized and controlled. The aquatic ecosystems are among the first to feel the effects of urbanization and agriculture. If the current United States population of 250 million continues to grow, and reaches the predicted 300 million by the end of the century, all current gains made in freshwater environmental preservation will be negated as the physical infrastructure is expanded to accommodate these added millions of people.

Contrary to conventional wisdom and wishful thinking, permanent environmental protection *cannot* be reconciled with economic development. Certainly the aquatic biological diversity that existed in pre-Columbian times can never be recovered. Many river systems and, indeed, entire drainages have been damaged beyond any hopes of recovery, and as development and growth continue unabated, aquatic environmental degradation can only worsen. A finite continental aquatic ecosystem cannot continue to absorb and recycle the wastes and effluent generated by a steadily increasing human population. As long as the global hydrologic cycle continues to function, the rivers and streams of America will continue to flow in the early decades of the next century, but they will be populated by fewer fish species of greater hardiness and adaptability.

Freshwater habitats, lacking the sea's vast cleansing and regeneration mechanisms, are at great risk due to their susceptibility to degradation or destruction by land-based human activities. But even the seemingly limitless seas are under assault in the latter years of the 20th century.

The major threats to the seas originate in the land-based activities of humanity and in the high-tech rush to harvest the oceans of their abundant protein. In the eternal hydrologic cycle, all waters that fall on land masses in the form of precipitation sooner or later end up in the seas—in one condition or another. Today, an increasing volume of the flow from land to sea carries an incredible array of toxic chemical and organic refuse, the byproducts of 20th century civilization.

The Hudson River, for example, drains an 800 square-mile watershed that includes several large cities. All these cities discharge volumes of municipal wastes of varying degrees of treatment into the river. New York City alone vents more than 40 million gallons of partially treated sewage each day into its own harbor. The Mediterranean Sea is subject to constant rivers of filth that have left southern Europe with the highest levels of typhoid fever on that continent.

Other disturbing trends—unknown a mere 25 years ago— include PCBs accumulating in the tissues of food fishes, lesions and tumors on fishes and sea turtles, gigantic toxic algae blooms triggering amnesic shellfish poisonings in people, and mysterious die-offs and mass strandings of whales, dolphins, and seals in both this country and in Europe. Further signs of a deteriorating marine biosphere include the death in the 1980s of nearly half of Europe's harbor seal population due to a viral infection; the destruction of vast shoals of menhaden by ulcerative mycosis, a disease first classified in 1978; and the accelerating destruction of coastal coral reefs through land-based erosion siltation and industrial runoff. These woes don't even take into consideration the avalanche of solid waste that enters the marine environments each year—everything from discarded shipping containers to plastic cans, bottles, disposable diapers, syringes, and six-pack holders form oceanic slicks miles long, ending up on isolated beaches or in the intestinal tracts of sea turtles and marine mammals.

Oil pollution of the seas is also a growing problem. The dismal litany of oil spill accidents takes on special significance when one learns that oil in concentrations as small as .005 parts per billion of seawater (about equal to five drops of oil in an average backyard swimming pool) is toxic enough to kill fish eggs and affect a fish's ability to move about and find food.

The very chemical composition of the seas is subtly yet inexorably changing due to the introduction of organically-rich fertilizers and acid rain from polluted rivers. Fresh water on land is diverted from the natural flow for the purpose of industry and irrigation and urban development, affecting the vast and complex marine food web in as yet undetermined ways.

Smaller, marine nongame or commercially important food fishes are at less risk of extinction than their freshwater brethren, simply because many of their ranges are much more extensive, and the sea itself has much greater recuperative powers. But the biodiversity of the fragile and sensitive littoral and continental shelf zones—those life-rich areas within 200 miles of the world's coastlines—will almost certainly decline further in the face of continued pollution assault.

The future is especially grim for many oceangoing species of commercially important fishes. At least 14 species have been so drastically reduced through relentless overfishing that even if fishing for them were to cease today the stocks could take 20 years to recover. Chief among these species are groundfishes such as cod, haddock, and flatfishes; red snapper; swordfish; bluefin tuna; and Pacific salmon. In the case of the tuna, the western Atlantic stocks fell from an estimated 300,000 individuals in 1970 to about 30,000 in 1990—a decline of more than 90 percent. The commercial incentive for the destruction of this beautiful animal is the astronomical price the Japanese market is willing to pay for them—up to $30,000 for a single fish in the Tokyo fish market!

Off the coast of South Florida, the red snapper catch declined by 97 percent in just six years; swordfish catches have declined more than 70 percent between 1980 and 1990. In the meantime, the U.S. fishing fleet has increased dramatically. The New England trawler fleet alone grew from 590 boats in 1976 to more than 1000 in 1984.

One additional "sea stress" factor that has become more important over the past 25 years is the introduction and spread of exotic fish species. It is from this group that many of the survivors among the fishes will be found. At least 48 exotic species of fishes have become established in the inland waters of the United States and Canada. "Established" means that a species has adapted to a new environment to the point that it is successfully reproducing itself over several generations and is expanding its range. The term "exotic" does not only imply the colorful guppies and gouramis of aquarium hobbyists (though both these fishes are established and breeding in parts of North America today), but means any fish, coldwater or tropical, that is not native to a given environment. Thus the infamous sea lamprey, the bane of the Great Lakes fishery, can be considered an exotic fish where it occurs in the Lakes, because it was granted entrance into the huge complex via the Welland Ship canal in the late 1800s. So, too, are the drab and unattractive carp, walking catfish, and the oriental weather loach, all solidly entrenched wherever they have escaped or been liberated into the wild. Native North American fish species—vital and integral parts of their own natural ecosystems—can quickly transform into dangerous exotics if they are carelessly introduced into water systems in other parts of the continent, where the endemic fishes might or might not be able to compete with them for food or living space. The yellow perch, largemouth bass, pumpkinseed sunfish, and the golden shiner—survivors all—are among the more striking examples of species that have been transplanted as game or forage fishes with varying degrees of damage to the receiving habitats.

With very few exceptions, all exotic freshwater fish species have spread across the globe through the agency of man. Deliberate introductions have historically been made by both individuals and government agencies for the following reasons: liberation of unwanted aquarium pets; establishment of gamefish stocks; mosquito or aquatic weed control; and introduction of ornamental fishes into private lakes and ponds fed or emptied by surrounding river systems. Accidental introductions are those that involve the discarding or escape of bait fishes; escapes from tropical fish farms or aquaculture facilities; aquatic organisms arriving on ship hulls or in bilges; and exotic species entering aquatic systems via alterations in natural waterways.

An enlightening example of just how rapidly a foreign fish species can conquer new territory is the story of the rudd (*Scardinius erythropthalmus*), an attractive European minnow similar in appearance and habits to our own golden shiner. This adaptable fish was originally introduced in New York State in the 1970s through liberations and escapes via bait dealers and aquarium hobbyists. It remained in that state until about 1980. By 1984, however, the rudd had turned up in a few apparently established wild populations in Maine, and by 1986 it was reported being sold in bait stores in 16 states and collected in open waters of 11 of those states. The species is today intensively cultured in Arkansas and widely distributed throughout the central and eastern states as a baitfish; there is little reason to doubt that it will eventually spread throughout the eastern and southern states as individuals escape the hook or are simply released after a day's fishing. The rudd is not included here as a survivor, but there seems to be little doubt that it will soon qualify for that label!

The essentially tropical cichlids, nearly all of them popular aquarium fishes, have also colonized large areas in a relatively short time. Seventeen cichlid species have become established in North America, and many of them were intentional introductions made by state and local agencies in the interest of sportfishing. Most of the successful cichlid introductions were made in the warmer Southern and Western states, but the oscar (*Astronotus ocellatus*), a large, colorful tropical fish popular with aquarists and introduced into Florida as a gamefish, has established breeding populations in such widespread and wintry places as Massachusetts, Pennsylvania, Rhode Island, and Ontario! The latter are doubtless the result of "midnight dumpings" of unwanted aquarium pets.

In addition to those exotics already firmly established here and competing with native fishes, another 65 foreign species have been collected in the wild in North America but have apparently not yet gained a solid foothold—in other words, established viable breeding populations. Only time will tell whether such aquarium favorites as the zebra danio, the neon tetra, or the kissing gourami will be among the North American fish survivors of tomorrow!

Sea lamprey
Petromyzon marinus

The sea lamprey is an Atlantic Coast species occurring from Labrador to Florida and present in landlocked populations in the Great Lakes and several lakes in New York State. This lamprey is widely distributed along the Atlantic Coast of Europe and in the Mediterranean Sea. It is an anadromous fish, that is, it lives in the sea and spawns in fresh water. Several populations have become landlocked, living entirely in fresh water lakes. In the Great Lakes this fish is considered a major pest, having caused the virtual collapse of the whitefish and lake trout fisheries.

The sea lamprey gained access to the Great Lakes via the Welland Ship Canal in the late 1800s. Its subsequent destruction of the Lakes fisheries was immense and nearly total, although eradication

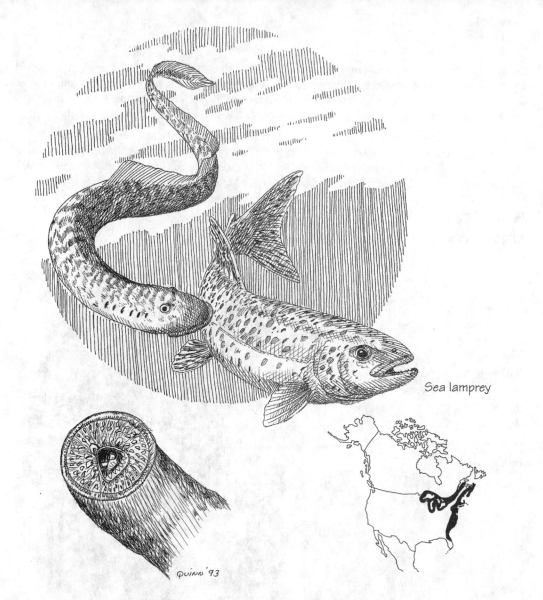

Sea lamprey

Quinn '93

measures carried out over the ensuing years have met with some success, bringing this fish's populations under some degree of control. In most cases the spawning streams are poisoned with a lamprey-specific fish poison that kills the young larvae, or ammocetes, as they emerge from the streambed gravels. The sea lamprey reaches the length of nearly 50 inches in the marine form; landlocked fish rarely reach lengths over 25 inches. This fish, and lampreys in general, are unattractive creatures. Primitive, eel-like fishes that lack bones and conventional jaws, lampreys are cartilaginous fishes that have a circular mouth opening filled with rasping teeth on the tongue and oral disk. Thus armed, the lamprey pursues other fishes, attaches itself to them, and rasps a hole in the unfortunate victim's flesh. The living tissues and body fluids of the host fish are consumed by the offensive parasite, which might or might not be shaken off after a period of feeding; in many cases the assault results in emaciation and eventual death. Many large fishes such as lake trout, whitefish, and ciscos have been caught with several lampreys firmly attached to them.

Although the sea lamprey is a pest of the first magnitude, most of the 18 other lamprey species found in the fresh waters of North America are nonparasitic as adults. The small brook lampreys of the genera *Ichthyomyzon* and *Lampetra* are secretive fishes that are seldom seen by the casual aquatic explorer. Averaging 6 to 15 inches, depending on species, the ammocetes of many of these lampreys remain hidden in the bottom mud and debris of quiet river backwaters and feed on small organisms, often for several years. On reaching maturity, the adults spawn and then die.

Although the sea lamprey is a tough, hardy creature well able to survive under degraded environmental conditions, the brook lampreys are as dependent on clean water as any trout, and soon decline and vanish when their streams are polluted with toxic runoff or silt. Several brook lamprey species are considered rare and declining due to habitat destruction.

American eel
Anguilla rostrata

Freshwater eels of the family *Anguillidae* are found on every continent except Antarctica and are familiar to almost everyone; people who have never seen a live eel at least know the common axiom, "slippery as an eel."

Eels are snakelike fishes that lack pelvic fins. Its scales are very small and the fish feels entirely smooth when handled. The color varies from a dark gray to olive-green and yellow. The head is small and the lower jaw has an underslung appearance. The American eel has a fascinating life history that remained a mystery until relatively recently, when it was found that the adults migrate from fresh water and coastal brackish habitats to the depths of the Atlantic Ocean in an area called the Sargasso Sea to spawn. Known as *catadromous spawning*, this behavior is opposite to that of *anadromous* fishes such as the various salmons, which spawn in streams and return to the sea to grow and reach maturity.

The eel begins life as a tiny, transparent, leaflike *leptocephalus*, which migrates with the prevailing currents across the wide ocean to the coast of North America, a trip that might take a year. On arrival in the coastal zones, the creatures transform into tiny "glass eels" that resemble the adult in appearance but are lacking in any pigment. From there, growth is rapid and proceeds into the "elver" stage, in which the fish become darkly pigmented but are yet of very small size—under three inches in length. The closely related European eel spawns in the same area, but returns to the rivers and lakes of the Eurasian landmass instead.

In the completion of the life cycle, the smaller males remain in coastal rivers and estuaries, while the larger females, which can reach the length of 48 inches, proceed inland, following rivers and then small streams sometimes for great distances.

The American eel is most abundant nearer the seacoasts. Here, in the lower reaches of virtually all eastern North American rivers and bays the fish forms the basis of an important and intensive recreational and commercial fishery, with hundreds of thousands of pounds taken yearly for both human consumption and the bait trade. Although the eel is not a popular seafood item in the United States or Canada—its snakelike form repels many people—it makes for delicious eating and has a dedicated following among Americans of black, Italian, and Eastern European ancestry.

This fish qualifies as survivor for two principal reasons: It is omnivorous, and it can survive in degraded waters that quickly kill other, less hardy fishes. Though considered a scavenger, the eel is an effective and determined hunter and can easily bag prey at night, when it is most active. Polluted rivers, bays, and harbors throughout the Northeast might lack many fish species, but all have large and seemingly thriving populations of common eels, and probably always will. New York Harbor and the nearby Raritan Bay in New Jersey have large eel populations and a sizeable eel fishery, though official warnings against the consumption of eels caught there are issued with dismal frequency.

The eel is capable of reaching a great age if all factors are in its favor. The oldest verified age for the nearly identical European eel (*Anguilla anguilla*), was that of an 88-year-old female in the public aquarium at the Halsingborg Museum in Sweden. This eel was reportedly collected in a river in 1860 as a three-year-old elver and died in 1948.

Bowfin
Amia calva

The bowfin is a big bruiser of a fish that looks mean and dangerous and for the most part it is, especially to any of its numerous prey. This fish is the sole surviving member of an ancient fish family that originated far back in the Jurassic era. Then, the group was widely distributed on all of the major continental landmasses; today it is restricted to North America, from the Saint Lawrence River and Great Lakes west to the Colorado River drainage and south to the Gulf States and Florida.

The bowfin is a powerfully built voracious predator that looks as though it belongs in a prehistoric swamp rather than at the end of a modern fisherman's line. Long-bodied and cylindrical in shape, the

Bowfin

creature has a large head and powerful jaws armed with many teeth. It is an ambush predator, lying in wait among dense weeds and debris and overtaking prey with a swift rush.

Normally of a mottled olive-greenish or brown color, the male becomes rather brightly colored during the breeding season. His mouth, throat, and belly turn brilliant turquoise green, and a prominent black spot on his tail is haloed in bright yellow or orange. He alone takes care of the many young, escorting them about and taking on all comers, regardless of size, who might do them harm. Large bowfins have been known to attack people who unwittingly waded too close to a flock of small fry.

This fish is nowhere overly abundant, which is just as well for all of the smaller fishes found within its range, but it is commonly found in swamps, sloughs, and river oxbows in standing or slow-moving water. It makes a spectacular strike at a lure or bait and fights with unbridled fury when hooked, but as the flesh is not considered palatable, it is not a popular food fish anywhere.

As befitting its primitive nature, the bowfin, like the African lungfish and the familiar Siamese fighting fish, has a lunglike gas bladder that allows it to take in atmospheric air. As such it can survive in deoxygenated and polluted waters that would prove fatal to other fishes.

Golden shiner
Notemigonus crysoleucas

A large and beautiful minnow, the shiner is one of the most popular bait fishes in North America. It has been widely and extensively cultivated and released in waters far from the perimeters of its original range. Naturally occurring on the Atlantic and Gulf Slope drainages from Nova Scotia to

Texas, and in the Mississippi River basin west to Montana and Wyoming, the golden shiner has been introduced in California and other places in the Far West, where it occurs in all habitats except higher mountain streams.

Golden shiner

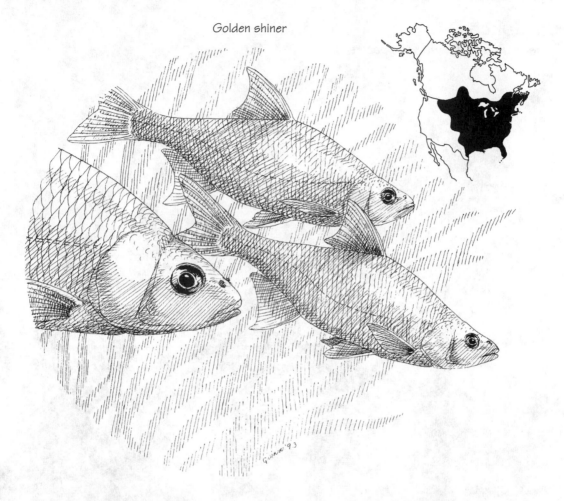

About a foot long, the fully adult breeding male is a sight to behold. Dark olive green above, its sides are a bright bronzy-gold fading to silver on the belly. Where the fish live in waters stained brown by plant tannins, the fins are dusky or reddish; in clear or murky waters they are unpigmented or yellowish. The scales of this shiner are often outlined in black, giving the fish's body a "chain-link" appearance. With a deep-bodied profile, a small head and mouth, and a strongly forked tail, this shiner is a graceful fish. It is highly active, and the sight of a small school darting and flashing in unison in the sunlight shallows is a small miracle of nature in itself!

The golden shiner has often been kept in aquaria by native fish hobbyists, but in confinement seldom attains the rich, bright colors of the wild fish. The shiner excels in the bait fish industry; hardy and tenacious, it survives long in the bait bucket. Indeed, it is through "bait bucket transport" that the shiner, along with several other species of bait fishes, has extended its considerable range. Bait species are often shipped far and wide over the nation to supply the angling demand, and many anglers simply dump any leftover fish into the lake or stream when day is done. The majority of fishes from individual dumpings do not survive the overall trauma, much less begin to breed, but there are millions of dedicated freshwater anglers today, and because this regrettable practice is widespread, the cumulative effect is felt in the establishment of a popular bait species far outside its natural range.

Catfishes
Ameiurus and Ictalurus species

Bullhead catfish

The bullhead catfish family is a large one, with about 40 North American species. Worldwide, there are 30 other catfish families containing over 2200 species. There are some eight species of Ameiurus, or true bullhead catfishes, also collectively called horned pouts. The three more common and widely distributed species are the black bullhead (A. melas), the yellow bullhead (A. natalis) and the brown bullhead (A. nebulosus), all occurring throughout the eastern and central United States and southern Canada, and widely introduced elsewhere.

Most North American catfishes, with the exception of the stream-living madtoms, are extremely hardy and adaptable creatures that occupy a wide range of habitats and rank high as fish survivors.

But even among this tough group the bullheads are exceptionally resilient. Some years ago an angler friend, knowing that I enjoyed fried catfish, left a creel full of freshly caught horned pouts on my back porch one evening, not realizing that I was away from home. Returning the next day I found the nine hapless fish, nearly dried out but still alive after spending at least 20 hours in the wicker creel without a drop of water. All of the fish recovered fully. They were placed in a sink filled with tap water and were later released, surely deserving their liberty after surviving such a prolonged ordeal!

Though the average bullhead is not often called upon to meet such daunting odds, the group as a whole is well-equipped for survival. These catfishes are prolific and take the parenting process a step beyond that of most fishes—they practice highly organized "child care." When the time comes to raise a family, the male clears a nest among the submerged vegetation near shore. The female deposits up to several hundred eggs in the nest, which is maintained and guarded by both parents. When the tiny blackish fry hatch, they are perfect replicas in miniature of their parents, and are guarded with unconditional commitment.

The young begin to leave the nest in about a week, but their departure is not an every-fish-for-himself affair. Instead, it's an orderly escort into the wonderful—and dangerous—world beyond. The adult cats herd the rambunctious fry into a tight school and shepherd them about for another week or two after they become "free swimming." Wherever bullheads are abundant, the milling swarms of tiny catfish slowly moving through the warm shallows near shore are a common sight from May through June. The little ones are allowed to investigate the terrain to their hearts' content—as long as they stay within the school's parameters. If the assemblage is threatened, the parents move in at once and ward off any attacker with fierce determination.

Bullheads are still quite abundant throughout their respective ranges, and although they are as affected by the acidification of their waters as any other fishes, they are among the last to disappear from northern lakes and ponds suffering from acid precipitation. Further south, these catfishes are in little danger of disappearing since they can easily endure degraded water conditions and have little trouble keeping ahead of the intense and relentless angling pressure to which they are subjected.

The blue catfish (*Ictalurus furcatus*) and the channel catfish (*I. punctatus*) are two American catfish species that attain a large size and are extensively cultivated for the "seafood" market. The channel cat has been widely introduced outside its natural range. Because it is a rather graceful and attractively colored creature (for a catfish), it has achieved some degree of popularity among aquarium hobbyists.

Walking catfish
Clarias batrachus

A native of Sri Lanka, India, and the Malay Archipelago, this extremely adaptable and hardy creature has become well established in peninsular Florida, and has been reported in the wild in several western and northern states. This rather unattractive though novel-looking catfish gained a foothold in Florida in the late 1960s after a number of specimens escaped the holding ponds of a Boca Raton tropical fish importer in 1964. It has since spread widely throughout the state, becoming established in creeks, ponds, and shallow lakes with weedy or muddy bottoms. Although its principal means of movement involves conventional swimming by the aid of fins, this fish is famous for its ability to move overland, either in search of new habitats or to escape from drying bodies of water. It locomotes by hitching itself along on its stout pectoral spines while briskly flexing its body.

This species is one of the labyrinth, or air-breathing, catfishes. A unique organ containing modified gill filaments supported by rigid, treelike structures allows the fish to take in atmospheric air and thus leave the water for extended periods on damp or rainy nights. Although the popular image of this fish is often that of a vaguely sinister, whiskered creature moving about at will by night and invading suburban gardens, in reality the fish attempts overland forays only when population pressures in smaller ponds force some individuals to move on, or when a body of water dries up. Under normal conditions they stay put, or move from place to place the way other fishes do—by following streams and rivers into fresh territory.

Walking catfish

The walking cat has been collected in the wild in California, Georgia, Nevada, and incredibly enough, in Massachusetts, although the decidedly tropical fish has virtually no hope of establishing itself in that state. Given its aggressive and territorial nature, I expect the walking catfish is here to stay and might well expand its range throughout most of the southern states in the future.

The walking catfish is about 24 inches long in its native habitats, but seldom exceeds 14 inches in the United States. It has been reported here both in the normal olive to brownish wild coloration and in the yellowish or *zanthic* and albino forms.

Goldfish
Carassius auratus

Native to Asia, the goldfish was first domesticated by the Chinese about 1200 years ago. In spite of the booming popularity of tropical fish, the goldfish is still the number one aquarium fish throughout the Western world, and without a doubt the domestic fish most familiar to people everywhere. Although the ordinary goldfish or "comet" of the pet trade is attractive in its own right, selective breeding has produced an incredible array of fancy goldfish varieties, some of them beautiful, some bizarre, and a few, such as the hapless bubble eye, grotesque.

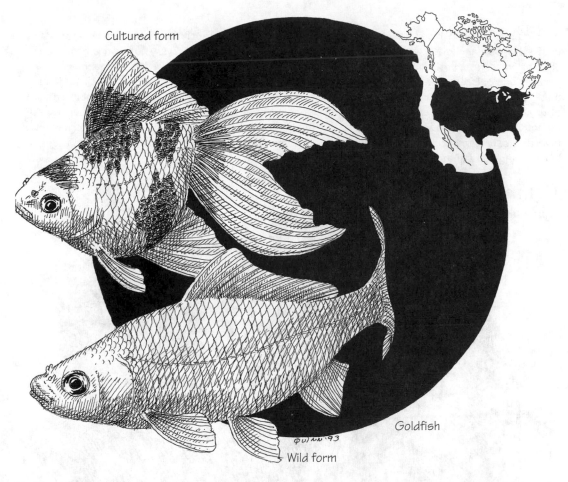

Cultured form

Goldfish

Wild form

Exotic goldfish strains such as the oranda, the lionhead, the various fancy fantails, Shubunkins and black Moors, to name just a few, are well-known to goldfish hobbyists. Such overbred, long-finned and big-headed unfortunates could not possibly survive for long in the natural environment, but the original wild goldfish is quite another matter. It is a hardy, savvy little fish able to survive adverse conditions in a wide variety of habitats. Like the carp, however, it has a decided preference for quiet, weed-chocked waters with soft mud bottoms for rooting. Goldfish, especially those recently released into the wild, do best in turbid waters that help conceal them from herons, kingfishers, and other fish-eating birds that would quickly spot them in clear water.

The goldfish was brought to Europe sometime in the last years of the 17th century; by about 1785 goldfish bowls were familiar household furnishings in England. In the mid-1800s, the fish found its way to the New World, where it quickly became popular as a parlor pet. A few were discovered in 1858 living wild in the Schuylkill River near Philadelphia, probably the result of deliberate releases of unwanted pets.

The intentional liberation of pet goldfish has been the principal agency for the spread of this fish, because it has never been very popular as a bait or forage fish. Through these releases, made all over the nation where the fish has been kept as a pet, the goldfish is now well established on the continent from southern Canada south into Mexico and breeding feral populations exist throughout Europe and in Hawaii, Australia, Madagascar, and most of the Orient. The goldfish hybridizes freely with the common carp, especially in areas of extreme pollution; the offspring are fertile.

Both goldfish and carp can attain ages in excess of 50 years in captivity, though few will ever attain that advanced age in the wild. Recently liberated goldfish are, of course, very noticeable because of their bright colors, but if the pioneers manage to survive long enough to breed, subsequent generations quickly assume the more somber, olive and bronzy-gold of the true wild stock, and are much better equipped for escaping predation by natural enemies. Goldfish can survive short periods of drying and entire winters frozen in the ice of ponds as long as the fish's tissues and vital organs do not freeze solid. Smaller than the closely related carp, an adult goldfish can reach 18 inches under favorable conditions. Because they are rarely as numerous as carp in a given body of water, they do less damage to aquatic vegetation or benthic organisms.

Common carp
Cyprinus carpio

Carp

Native to Eurasia, the common carp was first introduced on a very small scale into the United States beginning in 1831. The fish was successfully cultured in small numbers in New York throughout the following decade, but it was not until 1877 that wholesale distribution and release of the fish in our waters began in earnest. In the spring of that year, Rudolph Hessel, a government fish culturist seeking the perfect food and game fish for North America, imported 345 large carp that had been collected from the region around Frankfurt, Germany. Of that first fateful consignment, 220 of the fish were of the partially scaled, "mirror" variety; the rest were the fully scaled wild form that is found throughout the continent today. Hessel's carp were taken to Boston where they were placed in several ponds in the city's Druid Hill Park. Within two years the fish were crowding themselves, so arrangements were made to house some of them in the Babcock Lakes in Washington, D.C.'s Monument Park. Sixty-five mirror carp and 48 of the scaled variety were transferred to Washington, and the carp was on its way as a perfectly legal—and desirable—alien.

By 1879, the promotion of the carp as a superior food and sport fish was well underway. Their distribution became a political issue, with congressmen vying for supplies of breeding stock of the new "wonder fish" for their own districts. In 1879 12,265 were distributed to 25 states and territories. In 1883 more than 260,000 carp were distributed among 298 congressional districts.

For nearly a decade, the carp appeared to behave itself, and all reports on its progress were glowing ones. The big fish were fast breeders, and tame to the point of pethood. Nobody noticed any changes—good or bad—in the habitats into which the carp were dumped, so introductions continued right up until the turn of the century. Nobody had yet sampled the carp as a meal, either.

The first signs of trouble came when fishermen noticed that the fishing in "carp lakes" had declined and not improved, as the carp lovers had assured them it would. Nor did the carp prove to be the esteemed food fish here that it was in the Old World, and the market for the fish's flesh soon leveled off to its current relatively insignificant level.

In the years that followed the realization that carp were something less than a blessing, numerous and sometimes large-scale attempts have been made to eradicate the fish from the various waters it inhabits nationwide. Some of these campaigns have been successful, but they almost always involve poisoning entire lakes and ponds to remove the carp, because the wily fish is usually too smart to be easily seined or other wise caught by mechanical means. Intensive and thorough poisoning of a lake results in the complete elimination of the desirable game and forage species as well. The lakes must then be allowed to recover and then restocked with the desired native species.

The end result of the carp saga is that the fish is now, despite all attempts to control or eradicate it, widely distributed throughout temperate North America. It prefers rivers, lakes, and ponds with heavily vegetated and weedy substrates. It dislikes clear, cool, fast-moving waters and seldom becomes very numerous in such habitats. In most intact and unpolluted habitats distant from urban centers the fish maintains lower population levels, whereas in degraded waters they often swarm in great concentrations composed of hordes of stunted but mature fish of breeding age.

The recently introduced grass carp (Ctenopharyngodon idella) is an unrelated minnow of large dimensions that is showing all the signs of becoming an environmental nuisance. Native to Asia, this big fish (up to 50 inches) was introduced into Arkansas waters in the 1960s as a weed-control agent. Under a continuing program of weed control, it is now found, in 34 states, primarily through the agency of man. In one reported incident in Arkansas, however, several large grass carp among many kept in a diked holding pond were observed to leap across the narrow dike and into a natural waterway on the other side, successfully escaping into the local aquatic ecosystem!

Despite the fact that most of the fish liberated in official weed control projects are triploid—sterile individuals incapable of reproducing—the fish appears to be increasing slowly but steadily in the lower Mississippi River drainage. Time alone will tell whether the introduction of this hefty and voracious vegetarian will prove to be an environmental boon or bust.

Pumpkinseed and bluegill sunfishes
Lepomis species

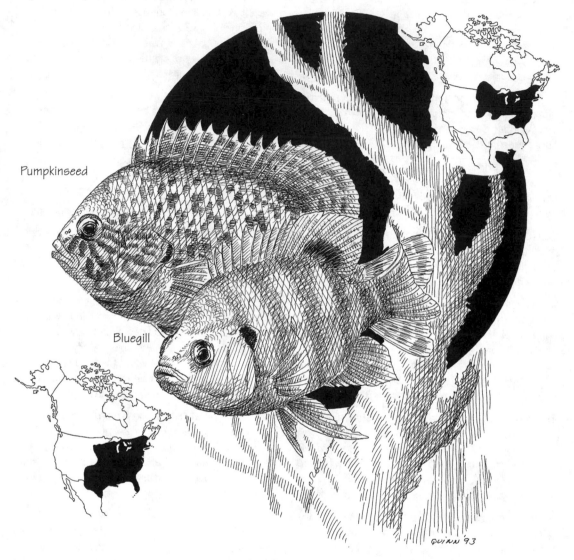

Pumpkinseed

Bluegill

It used to be said that the sunfishes, or "sunnies," as they're collectively called, were the fishes best known to schoolboys over the length and breadth of the land as the fighten'est little fishes to be caught on hook and line. The observation doubtless originated in those bucolic days before TV and video games, when the majority of schoolboys actually went fishing on a lazy summer afternoon. Whether or not they are still appreciated by the youth of America, the sunfishes are very much with us, and well known to everyone who does go fishing, young or old.

There are some 30 species of sunfish and basses in North America. It is an exclusively North American family of fishes, some of whose members have been widely introduced elsewhere in the world, including Europe, Australia, and Africa.

The majority of the sunfishes, in particular those of the genus *Lepomis*, are hardy and adaptable animals, varying from mildly to highly aggressive, like the tropical cichlids of the aquarium. Like the cichlids, sunfishes practice extended brood care of the eggs and young, thus reducing fry mortality and increasing the survival rate. For this reason, along with the fishes' ability to survive and even thrive under less than optimal environmental conditions, sunfishes are among the most abundant and least threatened of freshwater fishes today.

Without a doubt the two most familiar and widespread of the *Lepomis* sunfishes are the pumpkinseed (*L. gibbosus*) and the bluegill (*L. macrochirus*). Both are immensely popular panfishes (so-called because of the deep "pan-shaped" body, not because they will fit in a frying pan) that have been introduced far outside their natural ranges in the interest of sportfishing.

The pumpkinseed, or common sunfish, is everyone's idea of a typical sunnie. Originally native to the eastern United States from New Brunswick to South Carolina and west to North Dakota and Kentucky, this fish has been introduced to virtually all parts of the country and southern Canada. It has been widely stocked in private impoundments and farm ponds, and has found its way overseas where it is now common in many European rivers.

In form, the pumpkinseed does resemble that hefty vegetable's seed; the body is deep, rounded, and oval in profile, and might reach the length of 16 inches, though that size is exceptional today. This sunfish, like the mallard duck and the mummichog, is very common but strikingly beautiful at the same time. A breeding male, with his bright blue-green facial bars, orange body spots and yellow belly rivals many tropical fishes in sheer elegance.

The bluegill is a somewhat less gaudy sunfish but it is much more popular as a game and food fish, so much so that the species has been introduced all over the continental United States as well as into Hawaii and Europe. A really large bluegill might reach 17 inches, though an angler bagging one of that size is justified in heading straight for the taxidermist.

The bluegill's colors are on the somber side but it is nonetheless an attractive creature. It is olive-green above and iridescent yellowish green on the sides. Most large adults have a bluish sheen on the body and the breeding male has black pelvic fins. This sunfish has the elongated black "ear flap" on the gill cover, an anatomical feature shared by most sunfishes.

Sunfishes are common breeders in larger lakes and ponds and are often abundant in smaller city and suburban park lakes. The conspicuous nests, shallow, dishlike depressions dug in the sand bottom by the males in shallow water, are seen in spring and early summer. The male fish oversees the hatching of the eggs, stationing himself above the nest and vigorously fanning the eggs with his fins to keep them clear of silt and other debris. The young fry are protected from predators until they become free-swimming in about a week, after which they disperse and strike out on their own.

Male basses and sunfishes are quite aggressive when guarding eggs or young, and will readily strike a lure or bait cast near them. Anglers should refrain from catching such a guardian for the unprotected eggs or fry will quickly succumb to suffocation by silt or to predation.

Largemouth bass
Micropterus salmoides

Both this species and the smallmouth bass (*M. dolomeiu*) are centrarchids, close relatives of the much smaller sunfishes. There are about seven species in the genus *Micropterus*, all of them popular game fishes. The smallmouth bass occurs in more northerly areas where it frequents clear clean rivers and streams and the rocky shorelines of cooler lakes. It is a highly esteemed game fish, but because it does not attain the size of the largemouth and is less amenable to waters of poor quality, the fish has

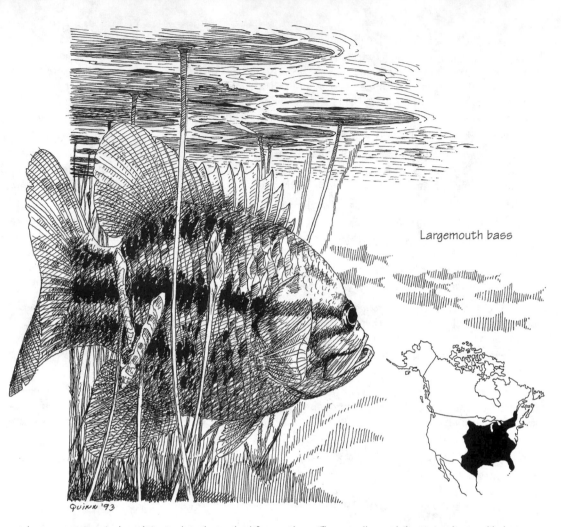

Largemouth bass

Quinn '93

not been as extensively cultivated and stocked for anglers. The smallmouth bass is thus unlikely to expand its range or even hold its own in the face of increasing habitat degradation in the future.

The largemouth bass, however, is another matter. This fish is the basis of a huge freshwater sportfishing industry and a dedicated angling fraternity. It is the object of countless bass derbies featuring millions of dollars in prize money held all over the nation as well as in Hawaii, where the largemouth has been introduced.

The natural range of this big, splendid gamefish includes the Saint Lawrence and Great Lakes drainages west to Minnesota and south to Florida and Mexico. It has now been introduced into West Coast waters, southern Canada, and even overseas, where it is established throughout much of western Europe and in parts of Australia. Its habitat preference differs from that of the smallmouth in that this fish does best in clear, well-vegetated, warm lakes and ponds and in the slow-moving backwaters of large and small rivers. Some of the largest bass—up to about 38 inches—will be found in artificial impoundments. The largemouth does not object to living in turbid, poor-quality waters, conditions that soon cause the smallmouth to decline and disappear. This environmental flexibility, coupled with its responsiveness to artificial culturing and stocking, should ensure that the various hardy strains of the largemouth will be with us for the foreseeable future.

Brown trout
Salmo trutta

The hefty, hearty brown trout is another Eurasian transplant that has made itself right at home in North America, as well as in several other far-flung places in the world. The big brash brown arrived in North America in 1882, with the object of introducing what was considered a premier game fish in Europe to the anglers of the New World. But long before that, the fish had traveled—by ocean-going steamer—to Tasmania, and then New Zealand and Australia, in a series of widely publicized and finally successful transplants.

Beginning with several ill-fated shipments of about 1000 brown trout eggs each beginning in 1851, repeated but botched efforts were made to bring the fish to the clear mountain streams of Tasmania in the 19th century. Finally, in 1864, 300 eggs arrived safely via the steamer *Norfolk* and were successfully incubated at a local fish hatchery. The fry were carefully reared and then stocked in local streams, where they did amazingly well in the absence of predators and spread like wildfire. Some of these initial fish were kept as brood stock and these formed the nucleus of the thousands of fish that were later shipped to parts far and wide throughout New Zealand and Australia. Today, New Zealand in particular enjoys the reputation of being one of the finest brown trout fishing destinations in the world; its rushing rivers and clear lakes are reportedly filled from bank to bank with lunker-sized browns along with rainbow trout imported later. The trout grow at the rate of about three and a half pounds per year per fish, on average.

The brown trout came to America almost two decades later, in two separate egg shipments from two separate Old World locations. In the winter of 1884, a German fisheries biologist named Von Behr sent some 40,000 trout eggs to one Fred Mather, superintendent of the Northville Fish Hatchery in New York State. Fry from this shipment were released into Michigan's Pere Marquette River the following April. Twenty-eight thousand fingerlings hatched from a second Von Behr shipment in 1885, and were introduced into several lakes and streams on Long Island, not far from Manhattan.

Periodic plantings of the fish went on until about the mid-1930s, and the fish is still stocked by the hundreds of thousands for the annual "put-and-take" trout fishing rodeo in almost every state of the union—even in streams too warm or polluted to support them on a permanent basis. Hailing originally from Iceland and Norway to the Mediterranean and Corsica, Sardinia, and Algeria, the brown is today a model citizen of waterways throughout temperate North America, as well those of New Zealand, Tasmania, Australia, Argentina, South Africa, and Sri Lanka, among other places.

The brown trout's claim to a position in the ranks of animal survivors stems from those two qualities common to all survivors: inherent hardiness and an aggressive temperament. The principal complaint lodged against the brown when it first made its debut here was that it was a coarse, "rough" fish that would quickly empty the streams of their rightful inhabitants, the lordly brook trout. This predicted scenario has—and hasn't— happened, depending upon the location, but in any case, the brown is without a doubt a survivor and the brookie is not, and for reasons we'll soon uncover.

The brook trout quickly declines and vanishes when its natal streams are subjected to degradation by siltation and warmed by the removal of the forest cover as regions are logged or urbanized. By the time of the brown trout's arrival in North America, the native trout of many of the continent's pristine waterways, particularly those in the rapidly developing East, were either going or gone. Thus, the tough, warmth-tolerant European immigrant was the logical inheritor of these sadly suburbanized streams. The brown trout has established viable breeding populations in many streams that are not under extremely heavy fishing pressure. Given its tolerance of pollutants and higher water temperatures the species seems assured of at least a temporary place in the fauna of the degraded stream habitats of the next century.

Mosquito fish
Gambusia affinis

The mosquito fish is one of the smallest yet toughest fishes known to science. Also called the potgut minnow, pusselgut (in Florida) and, simply, gambusia, this aggressive little creature has proven its mettle in its impressive ability to survive in all kinds of degraded habitats and among larger fishes that would gladly eat it if they could catch it. A large female mosquito fish will stretch the tape at about two inches; her mate can be considered a lunker at about an inch-and-a-quarter.

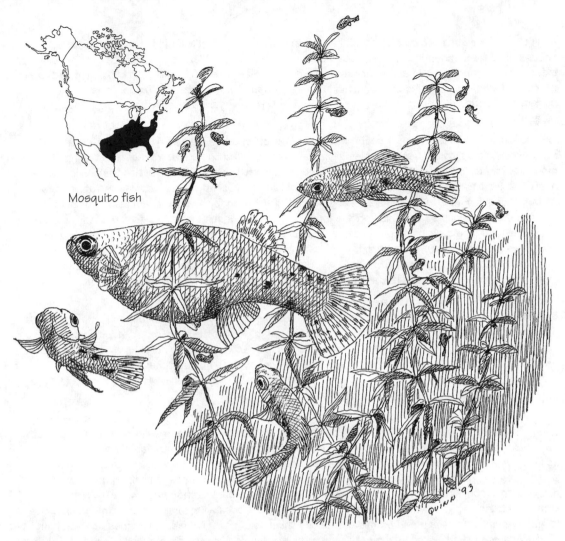

Mosquito fish

But in spite of the fish's lilliputian size, it has had a profound effect on human health and population growth worldwide, and makes its presence known in every habitat into which it is introduced.

Scientifically, the gambusia is a livebearing topminnow, closely related to the familiar and popular guppy, platy, and swordtail of aquarium hobby fame. As with those species, the male mosquito fish is considerably smaller than his mate and is readily distinguished from her by his specialized anal fin, modified into a tubelike structure called the gonopodium and used for the transmission of sperm into the female's genital pore. The species utilizes delayed fertilization, so that a female might deliver up to five consecutive broods after a single contact with a male.

The gambusia has been introduced all over the world as an effective and generally successful mosquito control agent; it has been naturalized in more than 70 countries—most of them in warm, malaria-prone regions—making it the most widely distributed freshwater fish in the world. Native to North America, from New Jersey and Illinois south to Florida, Texas, and into northern Mexico, the mosquito fish has been established in such far-flung places as Hawaii, the Philippines, Ceylon, Guadalcanal, Celebes, Samoa, Tahiti, and in every continent except Australia. It has been transplanted to the Far West and, although it is essentially a tropical fish, cold-resistant strains have been found living free as far north as Chicago and Connecticut.

In all there are eight species of small fishes in the genus *Gambusia* found in the United States, though all but the mosquito fish are found only in Texas and New Mexico. Several of these species are extremely rare and classed as endangered; the Big Bend gambusia (*G. gaigei*), for example, is today restricted to a single artificial spring-fed pond in Big Bend National Park in Texas. Other critically endangered *Gambusia* species have been decimated through hybridization with the aggressive mosquito fish.

The mosquito fish has, without a doubt, had great impact on the relentless fight against malaria and other mosquito-vectored diseases. A single female was reported to have devoured more than 150 mosquito larvae, or "wrigglers", in less than 10 hours, and with that kind of appetite the fish is capable of eliminating the bulk of the winged pests in any habitat both occupy. Large populations of gambusia in a given, mosquito-infested area have largely eliminated the need for heavy-handed use of dangerous and toxic pesticides such as DDT. This fish has found favor as a mosquito control agent not only because it eats mosquito larvae, which most smaller fishes do, but because of its impressive survivorship qualities; few fishes are more adaptable to horrible habitats or more fecund than this little creature. These factors, coupled with the small size that allows it to enter waters too shallow for other would-be mosquito-eaters, have made the gambusia the hummer-destroying champ it most surely is.

The mosquito fish is not a long-lived creature in nature; few individuals, and these mostly the larger females, survive longer than 15 months at most. But the fish's abbreviated lifespan is more than made up for by its fecundity. In warmer regions the gambusia breeds year-round, adult females producing broods averaging 40 live young every three or four weeks. More northerly-living fish usually have time to produce only four or five broods before senility sets in, but considering that an astounding 315 fully developed fry were found inside of one 2½-inch female, whatever time this little creature has at its disposal seems plenty to keep a pond swarming with mosquito fish.

Ever since people first began sowing gambusia all over the place back in 1905, its praises have been sung with gusto. It was hard to question the wisdom of dumping such a small, adaptable, and useful fish (even though the name "gambusia" is from the Spanish meaning "worthless") in waters where it could do what it does best—wipe out the local mosquito crop. But ichthyologists and biologists soon found the dark side to the mosquito fish. The problem lies in the fish's temperament, which is not what one could call introverted. Small though it is, the mosquito fish has few peers where it comes to taking what it wants. Where it encounters fishes occupying similar niches in new habitats, it quickly dominates them and eventually causes them to suffer declines. Male mosquito fish fight furiously among themselves in competition for the females, often with fatal results, and the species is strongly cannibalistic. Only the presence of abundant vegetation in the little fish's preferred habitats prevent them from keeping their own population at zero growth.

In response to the state of New Jersey's current plans to stock gambusia as a mosquito control agent, R. Bruce Gebhardt of the North American Native Fishes Association says simply that the mosquito fish "is the most feared exterminator of American fish species" and will "totally dominate" habitats into which it is introduced.

Mummichog
Fundulus heteroclitus

The mummichog, or zebra killy, is one of 40 species of topminnows and killifishes inhabiting the salt, brackish, and fresh waters of North America. As a group, the killifishes are among the hardiest and most adaptable of fishes. Many species can be classed as fish survivors in that they remain abundant today despite widespread and often severe changes in their habitats.

Mummichog

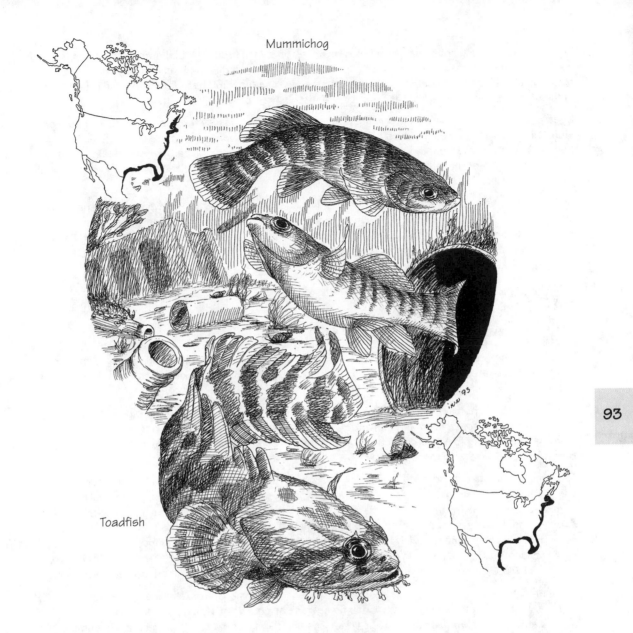

Toadfish

The mummichog, abundant and familiar to everyone who has spent time at coastal bays and beaches from Maine to the Gulf of Mexico, is typical of the group, and without a doubt a fish survivor. The fish's names reveal much about the creature's habits and habitat. *Mummichog* is a Lenni Lenape Indian term meaning "goes in crowds," which this fish most assuredly does; *killifish* stems from the Dutch word, "kill," meaning a creek or small river—the best place to look for this ubiquitous little fish.

The mummichog reaches the length of about five inches. In breeding trim the male is quite an attractive creature. During spawning activity he is darkish green above, silvery and rosy on the sides, with numerous reflective spots of green and gold. His underside and anal and caudal fins are bright saffron yellow, so that a crowd of lovesick males chasing a drab female is rather like a flock of swirling underwater butterflies.

The mummichog frequents shallow areas near shore, always in brackish tidal areas having dense aquatic or shoreside vegetation. It sometimes becomes stranded in mudflat pools by the receding tide, but this situation is seldom one of great peril, for the "mummi" has the ability to flip-flop over land to the nearest open water and safety—something few other fishes can do.

Used extensively as a bait fish and in both genetics and water pollution research, this fish has more than adequately demonstrated its tenacity under the most adverse circumstances. The fish are usually "dry-packed" when sold as bait; that is, they are placed in small cardboard containers with damp seaweed, a seemingly inhumane packing technique the fish nonetheless survive—even for a full day—with no ill effects.

Perhaps its greatest claim to fame is the mummichog's proven ability to survive and even thrive in the dirtiest water imaginable. They are abundant in the oily creeks and channels of the degraded marshlands around New York City, and are reportedly the only fish species able to survive and breed in the infamous Arthur Kill between New Jersey and Staten Island, considered close to the filthiest and most industrialized waterway in the world.

Oyster toadfish
Opsanus tau

The two species of toadfish found along the Atlantic and Gulf coasts are prime examples of estuarine fishes equipped both physiologically and temperamentally to survive in degraded marine habitats. The oyster toadfish occurs from Maine to Cuba in shallow brackish habitats; it is replaced in the Gulf of Mexico by the similar Gulf toadfish.

Both toadfishes are rather bizarre creatures, looking like oversized, mottled tadpoles with large mouths and a full complement of sharp teeth. They are ambush predators, lurking among weeds and debris and exploding forth to grab unwary prey. Toadfishes are creatures of shallow waters and are rarely found at greater depths offshore.

These fishes are able to tolerate the rapidly changing salinities and considerable levels of pollution of the estuarine environment. They can commonly be observed prowling about in extremely murky and turbid water near sewer outfalls. Toadfish are one of the few marine fishes that have not only adopted manmade artifacts as living and breeding shelters, but have come to prefer them over the natural objects they once used. Normally, the toadfish hides and spawns in empty bivalve shells or rock crevices, but with the regrettable increase in the volume of discarded trash in coastal waters, the fish now uses empty cans and bottles, pipes, old tires and other such trash as nest sites.

Toadfish lay relatively few but large eggs, secreting them underneath rocks or inside discarded objects. The male guards eggs and fry with fierce dedication. The guarding parent will not hesitate to attack any intruder, large or small, that threatens the safety of his progeny and many a summer wader has received a powerful bite on the ankle delivered by a big toadfish riding herd on the kids.

Toadfishes are capable of making sounds, particularly during the spawning season when the males can be heard giving vent to a curious, foghornlike noise from the depths in an attempt to attract a mate.

Cunner
Tautogolabrus adspersus

Among 20,000 known species of marine fishes, hundreds will doubtless exhibit some degree of the resilience and toughness required to survive the current and future assaults of overfishing and habitat degradation. Discussing even a small fraction of marine fish survivors—primarily those species which thrive today in coastal, brackish habitats worldwide—would require a hefty book in itself. The widespread cunner, a highly "intelligent" member of the tropical wrasse family, will serve as an apt illustration of survivorship among saltwater fishes.

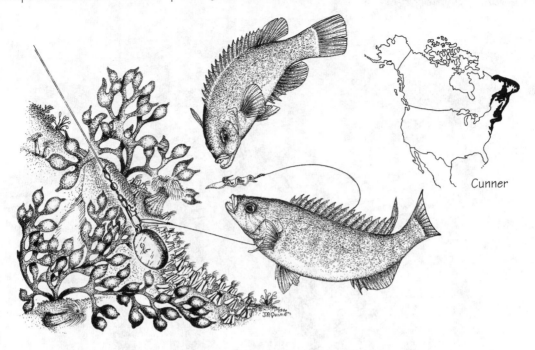

Cunner

The cunner, aka the choggie, chogset, sea perch, and bergall, reaches the length of about 15 inches in large individuals, though most are about half that size. The color varies considerably according to the prevailing habitat; jetty-bred cunners are usually a greenish brown with numerous red, green, and gold spangles, while those living among kelp and rockweed are bright reddish-brown. Numerous color intergrades exist, but all color phases have a white throat and chin, a clear identifying mark of the species. Like all wrasses, cunners have a long, spiny dorsal fin and conspicuous canine teeth in the jaws.

This small fish ranges from Labrador south to Delaware on the Atlantic Coast, where it is always found in rocky habitats or around dock and wharf pilings and seawalls. Although it was once considered a prime food fish, it is today regarded as a "trash fish" and bait-stealing pest among anglers, and is seldom eaten in any quantity by people. Its bait-stealing talents are among those very qualities that assign it survivorship rank, for the cunner is nobody's fool where it comes to avoiding capture by human fishermen or natural predators. Smaller fish can expertly strip a baited hook in seconds, and even large individuals easily avoid spearfishermen and sharks with ease. Although they swarm around popular fishing jetties and offshore wrecks up and down the Northeast Coast, relatively few are caught, and they far outnumber all other species found in the habitat.

The cunner appears to be one of the hardy marine species better able to resist the effects of coastal pollution, for few show up in the fish kills that periodically afflict the East Coast. The species is abundant even in harbors and inlets that are heavily polluted, and seems to be unaffected by the volume of non-point pollutants that wash into coastal areas following storms.

In general, cunners are more numerous and reach larger sizes at the southern portion of the range. This would indicate that the longer breeding season and the abundant and reliable food supply provided by the organic refuse and mildly polluted waters near U.S. urban centers have benefited the species.

Zebra mussel
Dreissena polymorpha

The discussion of aquatic invertebrates that will or will not survive into the next century and beyond would be a task of monumental scope. There are many thousands of described, and countless more undescribed, such organisms. The freshwater mussels, however, are a relatively large group of aquatic invertebrates that has attracted some attention in recent years, both with respect to the precarious state of native species and the recent appearance of an ominous invader.

Roughly 43 percent of North America's 300 native mussel species are currently classed as either extinct or very close to it. The great majority of these at-risk species are found in streams that have been dammed or polluted to the point that these specialized mollusks, unable to adapt to the new conditions, have dramatically declined and in some cases vanished entirely over the past 50 years. In particular jeopardy are the 60-odd species of mussels living in the vast Tennessee River system. A 1963 survey revealed that less than half of those mussel species survived in the Tennessee River and its tributaries. The cause of the decline was, for the most part, more than 30 dams obstructing the river system, creating a series of lakes rather than the healthy, flowing stream environment required by these bivalves for survival.

One mussel species under consideration for classification as endangered apparently slipped over the brink as scientists debated the issue. The winged maple leaf mussel, resident in the St. Croix River drainage in Minnesota and Wisconsin, was finally placed on the endangered species list in 1990, but by then, 99 percent of its stream habitat had been destroyed and the species was reduced to a single small population of the bivalves that was no longer reproducing itself.

As if all of these perils were not enough, American mussels face yet another threat, this one offered by one of their own—the imported zebra mussel.

The story of the invasion of North America by the zebra mussel is short, brutal and, for this continent's endangered aquatic environments, very ominous. This pint-sized, attractively striped mollusk suddenly appeared in the Great Lakes in the summer of 1988, an apparent stowaway in a transatlantic ship's bilge. It was first discovered in Lake St. Clair, a smaller body of water between Lakes Erie and Huron, but within two years the aquatic pest had overrun Lake Erie in concentrations as high as 700,000 mussels per square meter of substrate. The combined weight of the little mussels—individuals seldom exceed two inches in shell-length—was enough to sink boats and navigation buoys, and they quickly smothered native mussels by attaching themselves to the latter in huge numbers. Researchers found one beleaguered local mussel carrying 10,000 tiny zebra mussels on its overloaded shell, the hitchhikers stacked three deep in places.

The zebra mussel, native to the Caspian and Black Seas, had adapted to near-freshwater conditions there, and thus was able to make the move to North America with relative ease. The pest has already caused the shutdown of several municipal water supply plants on the Lakes, and has driven at least one native mussel, ironically named *Anodonta imbecillis*, to apparent extinction. Whereas this native bivalve once made up 38 percent of the mussel population near the Detroit Edison power plant, none were found there amidst a mass of zebra mussels in a survey made in 1991.

The urgent question among biologists today is, what's the next stop for the zebra mussel? Hopes that this molluscan version of the house sparrow would stay put in the Great Lakes were dashed when a large and apparently thriving population of zebra mussels was discovered in the Detroit River in July 1992, and smaller numbers were found in the lower stretches of the Cumberland and Tennessee Rivers. Today, apparently thriving populations of the zebra mussel have been reported in the Mississippi River from Minneapolis south to Vicksburg, the Erie Canal-Mohawk-Hudson rivers in New York state, and at isolated locations in the Arkansas and Susquehanna rivers. Biologists fear that the mussel will eventually colonize littoral hard-water habitats throughout North America.

The only hope, scientists say, is that the zebra mussel, which prefers quieter waters, will be unable to colonize the swiftly-flowing streams that are home to some of America's most endangered native freshwater bivalves. Since chemical and mechanical mussel controls are potentially damaging to the environment, biological control in the form of mussel-eating fishes, waterfowl, and crayfish might offer the best hope for staying the zebra mussel's advance across the continent.

But that hope is far from a certainty. In the meantime, the sinister zebra mussel, scourge of Great Lakes and points south, marches on—an aquatic survivor of the first magnitude.

Chapter five

The reptiles and amphibians

The accelerating decline of reptiles and amphibians worldwide is an environmental calamity of great concern. Historically, this animal group, with over 11,000 species—nearly twice as large as the mammals—has been viewed with suspicion and outright enmity by the majority of humankind. Few snake species, whether harmless and beneficial or not, escape direct persecution at the hand of man whenever contact occurs. Frogs and toads, possessing a vaguely human form, have fared somewhat better, finding some degree of popularity as novelty pets and, perhaps unfortunately, as laboratory and experimental animals.

A generally benevolent public perception can be a two-edged sword, however. The rising threat to the earth's reptiles and amphibians by the burgeoning demands of the pet trade is of increasing concern to biologists everywhere. In the mid-1980s it was estimated that in one year nearly two million reptiles and amphibians were collected for the pet industry in the U.S. alone. Worldwide, the total is much higher, with many millions of animals "harvested" yearly in the tropics by itinerant collectors and sent to pet shops all over the industrialized world, though primarily in the United States and western Europe.

The dramatic decline of frogs and some salamander species, particularly in North America, South and Central America, central and northern Europe, and Australia have been attributed to several factors, some obscure, others grimly definable. Principal among them are those inevitable old environmental nemeses, habitat alteration and destruction due to encroaching urbanization, followed closely by high-intensity agriculture and acid precipitation. The latter alters the water chemistry and destroys the algae crop so vital to the survival of these amphibians throughout the aquatic larval stage of their life cycle. Air and water pollution, pesticides, and direct persecution through collecting for food, laboratory use, and to satisfy the growing demand for off-beat pets completes the roster of grim forces arrayed against these creatures. Whatever the cause, many amphibian species have declined dramatically beginning in the late 1970s. This has occurred even in seemingly pristine habitats far from any obvious physical disturbance or pollution sources.

Reptiles and amphibians are hunted and collected for food in many parts of the world, but historically this type of exploitation has never been of a scope that might endanger entire populations. Exceptions might be the various species of marine turtles, which have come under increasingly heavy assault for both their meat and eggs in the past 50 years. Once isolated nesting beaches are now either easily accessible to subsistence hunters, or, as is the case in subtropical North America, now heavily developed and well populated year-round.

As noted above, two of the greatest threats to reptiles and amphibians worldwide are the laboratory animal and pet markets. The latter, in particular, is a tremendous drain on wild tropical reptile and amphibian populations, as well as certain temperate and American desert varieties. A study conducted in the early 1970s revealed that just under 600,000 amphibians were legally imported yearly into the United States for the pet trade; by 1988, that number had risen to nearly one million annually. Leopard frogs and garter snakes have long been popular experimental and dissection animals, and hundreds of thousands of these creatures were once collected for the biological supply house trade. Commerce in young sliders, cooters, and map turtles, during which millions were collected live throughout the South, has nearly faded away due to legislation banning their sale as carriers of the deadly salmonella virus, transmittable to humans. These turtles, however, are still exported to Europe in considerable numbers, where their sale is not prohibited.

Today, North American reptiles and amphibians, or "herptiles" in scientific jargon, are declining almost everywhere due to destruction and reduction of habitats to accommodate urban and suburban development. Because of their ground-hugging nature, reptiles and amphibians are highly susceptible to the fragmentation of their often fragile and specialized habitats. The network busy roads and freeways that invariably accompany development present insurmountable barriers to migration and free movement between habitats. Everyone who has driven suburban and country roads knows that snakes, lizards, and toads were once—but no more—among the most commonly spotted roadkill victims in the front lines of the "new suburbias."

Many species of frogs and salamanders appear to be on the decline, even in rural and near-wilderness areas where their disappearance can be less easily explained or attributed to obvious human-related causes. If present trends of human population growth and development continue, the future will see the further decline and eventual disappearance of all but the more adaptable and hardy species. The few cold-blooded vertebrate survivors will be those able to coexist with man, either through their canny ability to maintain a low profile or to adapt to the altered and degraded habitats and higher levels of environmental pollution that the other, more specialized forms found to be lethal.

The reptiles and amphibians discussed here are a sampling of those forms that have long occupied a high profile in the public perception simply because they have remained relatively common and noticeable, even in the environs of our largest cities. Snapping and painted turtles thrive in the smallest and most recreationalized of park ponds, while brown snakes and dusky salamanders can still be discovered going about their furtive ways in vacant lots and abandoned city landfills. Although we will probably not see the total demise of the species discussed here unless we completely destroy the marginal habitats they cling to today, nothing can be guaranteed for the futures of any reptile and amphibian species, with the possible exception of the snapping turtle, which can survive the most degraded of urbanized aquatic habitats with little trouble as long as a food supply exists.

Snapping turtle
Chelydra serpentina

This large, primarily aquatic turtle is one of the more widely distributed North American reptiles. Existing in four scientifically debatable subspecies, the snapper is found from Nova Scotia and Quebec west to southeastern Alberta and southward east of the Rockies to Florida and Texas. The range continues south through Mexico and Central America to Ecuador.

Snapping turtle

The American snapping turtle occupies nearly every conceivable freshwater habitat, including lakes and streams, sluggish rivers, impoundments, farm ponds, bogs, and even brackish water marshes. It is one of the relatively few reptiles that can readily tolerate seawater, and commonly enters bays and estuaries. (Sea turtles, sea snakes, and the diamondback terrapin are among reptiles specifically adapted to the marine environment.)

The snapper is something of a paradox: almost everyone has heard the name and has a good idea as to what a "snapping turtle" is, but very few people beyond dedicated fishermen or the minority of practicing herpetologists have ever met one in the flesh or have had meaningful contact with the species at all. Nevertheless, everybody knows that the snapper is a beast to be reckoned with . . . dangerously unpredictable, prone to unprovoked attacks upon wading children and lakeside picnickers, in short, a relentlessly evil, essentially useless creature if there ever was one.

Like all "animals with bad reputations," the snapping turtle's bad press is only partly deserved, and none of the few justifiable complaints lodged against it are the fault of the turtle in the least. A snapper will eat baby ducks and appropriate its share of fish from the human angler's catch because these are the turtle's food, along with anything else it can catch. A snapper will bite and hang on when provoked, but almost any normal creature—people included—will defend itself if molested. The snapper is considered foul-tempered and ugly in appearance, but so are many people, an opinion we'll pursue no further here.

The snapper's basic problem is that it is an ancient anachronism. Of the same species as the dinosaurs, the big, fearless, and heavily armored reptile is very much dinosaurian in mien, and thus has survived all that a hostile world can throw at it for eons. The turtle survives today in pretty much the same form that served it so well all those millions of years. An adult snapper, which can reach the weight of 60 pounds, has few, if any natural enemies. Its physical construction and outward appearance is that of total utility; gaudy, frivolous colors and delicate, wavy appendages are not a snapping turtle's stock in trade.

Snappers are very long-lived creatures, attaining the age of about 40 years in the wild, possibly much longer in captivity. A verified record concerns the closely related alligator snapping turtle, one of which died at the Philadelphia Zoo in 1949 at the age of 58 years and nine months. As this particular turtle was accidentally killed, there is no way of knowing how much longer it might have survived.

The snapper is omnivorous; depending on size, it will eat any animal it can overpower, up to and including snakes, frogs, birds, small mammals, and other turtles, including smaller individuals of its own kind. Where they are abundant, snappers have been implicated as serious predators of young waterfowl.

In spite of their size, ill-temper and appetite, snapping turtles are able to coexist with humankind due to their essentially secretive nature. Primarily nocturnal, the turtle basks and lazes about by day and is seldom seen. The male rarely voluntarily leaves the water, though the female must do so each year in order to excavate her nest and lay her white, spherical eggs, which can number between 15 and 75 in a large female. Snappers are inordinately aggressive and display a readiness to strike when molested on land. In the water they are normally docile, and make every effort to escape if bothered—though this statement is far from carved in stone! The bite of a snapper can be dangerous, and even a baby one will strike if handled. A captive snapper will never become a "pet" in the manner of the more docile painted and map turtles.

The related alligator snapping turtle (*Macroclemys temminckii*) occurs in the Mississippi Valley from Illinois south to the Gulf of Mexico and east and west to Florida and Texas. It attains much greater size than the common snapping turtle, with 200-pound individuals reported in the past. The species is considered the heaviest freshwater turtle in the world.

Painted turtles
Chrysemys species

One of the most familiar aquatic turtles, the painted turtle exists in four distinct subspecies from coast-to-coast and from Ontario and Nova Scotia south to the Gulf of Mexico. Throughout much of the northern part of this vast range they are the only conspicuous basking turtles. Groups of them are frequently seen sunning themselves on rocks and logs projecting from the surface of lakes and ponds, often piled one atop the other in their eagerness to soak up some sun.

The eastern painted turtle (*C. picta*) is best told by its dark, smooth, rounded shell, or carapace, and the red or yellow patterns at its edge. The head usually sports several fine yellow lines and two bright yellow spots on either side; these marks are highly conspicuous fieldmarks and are usually visible from some distance away. The undershell, or plastron, is a plain goldish-yellow, though it might show a few darkish spots or blotches. The sexes are pretty much alike in appearance, but the male will be brighter in color and pattern and has noticably longer nails on the front feet.

This four- to six-inch turtle is without a doubt the chelonian most familiar to Americans from coast to coast. The nearly identical midland painted turtle (*C. picta marginata*) is found from southern Quebec and Ontario south to Georgia and Alabama; in the west, the eastern painted turtle is replaced by the similar western painted turtle (*C. picta bellii*), which differs from the eastern form primarily in the intricate, netlike markings adorning the carapace.

Painted turtles

Painted turtles can be found in all manner of aquatic habitats, from lakes, marshes, and bogs in the east to prairie sloughs in the west. They frequent the quieter backwaters and pools of larger rivers and might venture into brackish water marshes near the coasts. During the spring breeding season these active turtles wander about on land and are among those most commonly seen crossing roads in search of a nest site. These turtles appear to be tolerant of moderate water pollution and physically altered habitats. They are common residents of suburban park ponds and cattle tanks. Once past the more vulnerable hatchling and juvenile stages, they have few enemies in the suburbanized environment and can usually evade capture by avoiding shorelines lacking in vegetation. The majority of large painted and map turtles are observed basking on logs and rocks some distance from shore.

Painted turtles have been caught and kept as pets by countless generations of American and Canadian kids, and are prominent among those relatively few reptiles that can be considered pets at all. The hatchlings, slightly larger than a quarter in shell size, are most appealing little creatures that tame quickly and readily approach their keeper for tidbits. These turtles can be kept in an "aquaterrarium" featuring a shelving sand beach and a few inches of water. They will readily accept virtually all meat and fish fare, as well as earthworms, mealworms, and live goldfish. The important requirements in turtle keeping are clean quarters and plenty of light and warmth. Maintained under ideal conditions, a hatchling painted turtle will live for 20 years or more in captivity.

Sliders and cooters
Trachemys and Pseudemys species

Sliders

Cooters

quin

The sliders are relatively large turtles commonly observed basking on logs or rocks in ponds and lakes throughout the southern portions of the United States, sometimes stacking themselves two and three deep in their eagerness to "get some sun." Sliders, so-named because they will vacate, or "slide" from a basking spot at the slightest indication of danger, were once immensely popular as children's pets. Millions of unfortunate hatchlings, their shells painted in gaudy designs and colors, were peddled in pet shops and carnivals all over the land until humane concerns coupled with fears of salmonella transmission ended that huge market.

The red-eared slider (*Trachemys scripta elegans*) is perhaps the best-known species of the group and has the widest distribution. Occurring naturally from West Virginia south to the Gulf of Mexico and east to New Mexico and northeastern Mexico, this turtle, due to its status as a widely kept pet, has been introduced in many places far outside the normal range for the species. The great majority of these smaller populations are the result of the liberation of unwanted pets, and populations of the red-ear in Maryland and Ohio probably had their genesis in such releases made repeatedly over a period of many years.

This turtle grows to a shell length of about eight inches; the most conspicuous fieldmark is the broad reddish stripe on the head just behind the eye, lending the turtle its common name. The stripe, which might be yellowish in some individuals, is unique to the sliders; no other group of American turtles has this feature, though in larger, darker, adult red-ears the stripes might be obscured or absent. The shell, or carapace, of smaller individuals is usually a rich olive-green with darker lines or vermiculations, while those of older animals are darker, becoming mottled brownish or even black. The underside, or plastron, of small sliders is a bright yellow, marked by brown or blackish eyelike spots.

The closely-related cooters (*Pseudemys* species) are generally larger than the sliders, reaching the shell lengths of 10 to 14 inches at maturity. The eight species of "river cooters" have been utilized as food for hundreds of years throughout their ranges and in the past were nearly as popular as the sliders as terrarium pets. These turtles are generally tolerant of degraded water conditions and habitats and remain common throughout most of their ranges. The name "cooter" is of African origin and stems from the word, kuta, a generic term for turtle in several African dialects; the word was brought to North America with the slave trade in the 18th century.

The hieroglyphic river cooter (*P. concinna hieroglyphica*) ranges from southern Illinois south to northern Mississippi and east to Tennessee and Georgia. As in most cooters, this turtle occurs in almost any aquatic habits offering softer, muddy bottoms, abundant vegetation, and plenty of rocks and logs for basking. The existence and location of basking sites in a cooter pond has important survival value, for these extremely wary turtles drop into the water at the slightest sign of danger and are very difficult to catch where they have a broad overview of their surroundings. This has doubtless contributed to the cooters' continued survival and abundance in an increasingly hostile, people-filled world!

Map turtles
Graptemys species

Among the more common North American freshwater turtles the various map turtles are probably borderline candidates for reptilian survivorship. Although several species remain abundant throughout their ranges, especially in the southern states, many populations have been decimated by the pollution or channelization of their aquatic habitats. In the past the hatchlings and young animals were intensively collected for the pet trade, a practice that has doubtless also contributed to map turtle decline. These attractive turtles are medium-sized chelonians, reaching a carapace length between 7 and 13 inches at maturity.

As a group the map turtles are widely distributed from southern Canada and the Great Lakes to the Gulf of Mexico, east of the Rockies. They are creatures of larger lakes and rivers, and most are addicted to prolonged periods of sunning on rocks, logs, and other projections above the water surface. These turtles are omnivores, eating a wide variety of plants and animals encountered in the habitat. As they strongly favor snails and mussels in the diet, however, they could be considered susceptible to decline as native freshwater snails and mussels disappear due to stream alteration and pollution.

The name "map" turtle stems from the often-intricate, maplike tracery of lines on the heads of both adults and young, and the decorative patterns adorning the shells of the hatchlings. In two species, the false map turtle (*G. pseudogeographica*) and the Mississippi map turtle (*G. kohni*), the iris of the eye is bright white, giving the creatures a distinct, bright-eyed staring look. The map turtles display strong sexual dimorphism in that the females are considerably larger than the males, sometimes by as much as two or three times; in most species this means that an average adult male might be about five or six inches long while his mate might attain 11 to 12 inches in shell length.

Map turtles

The false and Mississippi map turtles are still rather abundant in lakes and rivers in the South, though they have become rare in parts of the Mississippi River drainage below Kansas City and St. Louis due to habitat alteration and high levels of pollution. Although live trapping of the young for the pet trade has been outlawed, the adults are still caught for this purpose and a fair number find their way into retail pet stores today.

In the past the map turtles, in particular the false and Mississippi species, were immensely popular as pets due to their attractive colors and patterns. Although less hardy and adaptable than the sliders and painted turtles, these creatures were sold through pet shops in great numbers until fears of salmonella poisoning among human turtle owners resulted in the prohibition of the sale of all young turtles with a shell length of four inches or less. The more unfortunate aspect of this desire for unusual pets is that few of these appealing but rather delicate "baby turtles" sold through retail outlets to casual pet owners survived more than a few weeks or months in captivity. Less amenable to handling, with more specific requirements concerning foods, adequate light and warmth, and a deeper-water environment has always made keeping map turtles an enterprise better suited to the experienced reptile keeper.

The keeping of reptile and amphibian pets is fraught with environmental drawbacks. Exotic and difficult-to-keep creatures are imported or collected locally and sold through retail outlets throughout North America, all to satisfy the desire for "ownership" of the unusual among living things. I was employed in the wholesale tropical fish and herptile industry some years back, and still recall with sadness the volumes of such delicate creatures as spotted salamanders, anoles, newts, and various native and exotic frogs and toads—all of them subjected to rough handling through the break-neck process of collection, transport, and resale—destined to an early demise in a 10-gallon aquarium in the hands of a well-meaning but inexperienced pet owner.

Green anole
Anolis carolinensis

The anoles, with more than 250 species worldwide, are the largest genus of lizards in the world. Only one species, the green anole, is native to the United States, although several Caribbean and one southeast Asian species have been introduced and are established in southern Florida. The anole is perhaps the one lizard most familiar to the public in general and pet owners in particular, usually under the erroneous alias of chameleon. The green anole occurs naturally from North Carolina south to Key West and west to southeastern Oklahoma and Texas. These are highly active, strongly arboreal (tree-living) lizards that have adapted very well to mankind and his dwellings throughout the South. They are found in virtually all habitats from overgrown farms and rural districts to both the urban and suburban areas of Miami and other towns and cities in Florida.

These small lizards have long been popular as pets and many thousands are still collected in the wild and sold through retail pet shops. Although these creatures are hardy and long-lived in the wild state, they generally do not survive long in captivity unless given very good care. This includes providing enough natural or artificial light and warmth, plenty of live insect foods, and water on a daily basis—delivered by misting or sprinkling the terrarium's vegetation so the anoles can lick up the droplets. This animal's quarters should be securely covered, for if an anole escapes inside a home it can be quite difficult to recapture!

Garter snake
Thamnophis sirtalis

The garter snake is probably the most abundant and familiar serpent in North America. This snake occurs in 13 subspecies—one of them on the federal endangered species list—and is found from southern Canada throughout most of the United States and northern Mexico. In general, all garter snakes are dark brown, olive, or black in ground color, with a bright yellow, greenish or bluish dorsal stripe and two side stripes running the length of the body. There are usually a series of staggered dark brown or black spots running between the stripes, though these might be indistinct or merge with the lighter stripes.

The slender garter snake is an alert, intelligent creature well equipped for survival in the altered environment. Although it prefers moist wooded and shrubby areas near lakes, rivers, and bogs, this snake is often seen at field edges, golf courses, roadsides, and suburban lots, as long as there is enough cover and abundant prey. They are aggressive and persistent hunters, catching and eating all manner of small prey, from insects and earthworms to frogs, fishes, and salamanders. Unlike the slimmer and shyer ribbon snake, a closely related species, a big garter snake will vigorously defend itself when caught and handled, biting hard and releasing a foul-smelling musk from its anal glands. In spite of its feisty temperament and ready bite, this snake makes a good pet, if any reptile can be said to qualify for true pethood, and quickly becomes tame in captivity. Thoroughly acclimated animals will readily accept strips of fish or earthworms from their keepers' fingers and can be handled without fear of a bite.

The species has repeatedly bred in confinement, bearing between 14 and 40 live young which are about seven inches long at birth. The snakelets are fast-growing and can reach the length of 30 inches in two years, though this size is rather unusual for the species.

The only North American snake currently on the federal endangered species list is, in fact, one of the garter snakes—the critically endangered San Francisco garter snake (*T. sirtalis tetrataenia*).

Brown snake
Storeria dekayi

The little brown snake is the epitome of the harmless serpent. Seldom exceeding 15 inches in length, the animal is a slender, small-headed snake that will not bite even when plucked roughly from the ground and handled. Its sole defenses are to flatten its body in order to look larger, and to exude a mildly-odorous musk from its anal glands. A new captive hardly struggles at all; it seems to resign itself to its fate and can be handled as though it had always been a pet.

The brown snake is still a very common serpent, occupying a wide range from northeastern Canada and the eastern half of the United States south to Mexico and Honduras. The animal's qualifications for survivorship are largely due to its flexible habitat; it can be found anywhere there is suitable cover and smaller prey animals. Wet or dry locales, woodlands, meadows, swamps, and roadsides all suit it, and it can even be found in suburban parks and vacant, trash-littered lots in the centers of larger cities. Its ability to survive hinges on the availability of protective cover, whether dense undergrowth and rotten logs, or piles of refuse, cardboard, and sheet metal. Nocturnal in nature, the snake remains hidden during the day, which accounts for the fact that although this snake can exist within the shadow of skyscrapers in some cities, the vast majority of human residents are totally unaware of its presence.

Newborn Dekay's snakes are about three inches long and are dark brown or black with a white neck ring. They are delicate creatures, but they can be kept with relative ease in a terrarium and fed small earthworms, fruit flies, and meal worms. The closely related but less abundant red-bellied snake (*S. occipitomaculata*) generally occupies the same habitats as the brown snake. It is similar in coloration and form, but has a reddish or pink belly and seldom exceeds a foot in length. This little snake is also adapted to living in close proximity to humans, and can be found under refuse and trash in vacant lots and weedy open areas near cities.

Red-backed salamander
Plethodon cinereus

109

As an amphibian group, the salamanders and newts are probably the least known to the average person. Although numbering some 300 species, salamanders are not often observed by the casual nature explorer unless searched for specifically in their aquatic or woodland habitats. The eastern newt and the spotted salamander are two species that are both larger and more active, and thus come to mind when the average person thinks "salamander." The eastern newt can be found in both an aquatic adult stage and a juvenile, land-based form called the red eft. The six-inch spotted salamander assembles in often dense breeding aggregations in woodland pools in the early spring, though in recent years their numbers have declined due to habitat destruction and collecting for the pet trade.

Among amphibians in general and salamanders in particular, the red-backed salamander might qualify for survivor status based on several criteria. It is a very widely distributed little animal, and at three to four inches maximum, it is small and inconspicuous, and thus not overly appealing to amateur herp collectors. Finally, because it is a terrestrial salamander, it is not as dependent upon an unpolluted aquatic habitat as are most of the other North American forms. The little red-back has as good a chance of surviving as any amphibian, as long as some woodlands escape the ecological death of pavement.

This salamander occurs from the Maritime Provinces and southern Quebec to Minnesota and south to North Carolina. Throughout much of the South it is replaced by the similar southern red-backed salamander (*P. serratus*). The red-back can be found in nearly all wooded or forested areas throughout its wide range, hiding during the day underneath bark, leaf litter, rocks, and all sorts of trash and rubble. It emerges at night, especially during rains, to forage for the tiny insect prey it eats. Unlike many other amphibians, which deposit their eggs in water, the woodland salamanders lay their eggs in a clutch beneath damp leaves and other forest litter. The female encircles them with her body until the young—perfect miniatures of their parent—emerge from the eggs. This method of reproduction eliminates the need for a larval, or tadpole, stage of development, which requires an intact, unpolluted body of water for successful completion. Many amphibians that include a tadpole stage in their life cycle have been decimated in parts of the continent due to the acidification of their breeding lakes and ponds.

Bullfrog
Rana catesbiena

The deep, sonorous *jug-o-rum* of the calling bullfrog is characteristic of warm spring and summer nights on sloughs and bayous throughout the South, though this big amphibian is by no means restricted to parts south of the Mason Dixon Line. Occurring from Nova Scotia and Wisconsin south to Florida and Texas, the bullfrog has been introduced far outside its natural range, including California and British Columbia in the Far West, as well as Cuba, Jamaica, Mexico, and Great Britain.

A voracious predator, the bullfrog is a good candidate for pest status. Although a threat to endemic wildlife wherever it goes, it is never too abundant in any given locality, so its overall impact is not likely to be severe. Researchers have reported that where it becomes numerous in new habitats, it has an adverse effect on other frogs, small snakes and turtles, and the young of waterfowl. The bullfrog is a rather sedentary creature that functions as an ambush predator; that is, it spends relatively long periods of time lying in wait for prey, which it seizes in a sudden rush. Most of this big amphibian's time is spent hiding among the dense greenery of its aquatic home, though given its powerful leg muscles, a bullfrog can move if it has to. An example of its leaping prowess might be found in a bullfrog named Rosie the Ribeter, which won the broad-jumping event in the annual Calaveras (California) Jumping Jubilee in 1986 with a leap measured at 21 feet, 5¾ inches.

Bullfrog

Green frog

American toad

Green frog
Rana clamitans

Nearly all North American frog species have undergone declines in recent years. Although the green frog has not escaped this trend, it appears to be more abundant than other species, and better able to retain its hold in degraded or urbanized aquatic habitats. It is still the commonest frog to be found in much of the eastern and central parts of the continent today, and although local declines have been reported, this species appears likely to persist.

The leopard frog (*Rana pipiens*) is another widespread frog species that, while it remains abundant today, is much less common than it was a mere 25 years ago. This agile "meadow frog," once collected in the millions as a high school dissection and laboratory animal, is not as tied to the aquatic environment as are the green and bullfrogs; it can be found in a wide variety of upland habitats well away from water. The southern leopard frog (*R. utricularia*) regularly enters brackish marshes along the coast, though it cannot breed in such saline environments.

The leopard frog, perhaps more so than the green frog, is familiar to the average person due to its wide use as a scientific specimen and to its habit of turning up in suburban yards and gardens and other cultivated situations during the summer months. It is brown, tan, or greenish with two or three rows of irregular dark spots on the back, and rounded spots on the sides. In other words, it looks like everybody's idea of what a frog should look like. This creature has a very broad range, occurring virtually across the continent, from Quebec to north-central Canada, south to Kentucky in the East and Arizona and New Mexico in the West. The southern leopard frog is found throughout the southeastern states into the Florida peninsula.

The leopard frog is one of the more acrobatic frogs, making a series of swift, darting and erratic leaps when disturbed. Attempting to catch a leopard frog in a grassy field has to be one of the more exhausting and futile of outdoor experiences!

Spring peeper
Pseudacris crucifer

Many more people know the spring peeper by its voice than by sight, for few have ever seen this elusive little creature in its natural habitat. Although the ethereal spring voice of the peeper has been all but overwhelmed by the noises of suburban civilization in many areas—and stilled forever in others where its wooded wetland habitats have been destroyed—the tiny tree frog is still one of the most widely distributed of North American amphibians.

The peeper is found from the Canadian Maritime Provinces south to northern Florida and west to southeast Manitoba and Texas. It has even been introduced into Cuba. The peeper exists in two subspecies, the northern and southern varieties. The two are virtually identical save for the presence of blackish speckling on the belly of the southern spring peeper.

The spring peeper is close to the smallest North American frog, reaching a maximum length of one and a quarter inches at maturity. The color varies from pale yellow and brownish through gray or olivaceous, but the animal's back is always adorned with the irregular, cross-shaped X-mark that inspired its specific name crucifer, meaning, "cross-bearer." The peeper is technically a kind of tree frog, but it rarely ascends plants to any great height and then only during the breeding season, when it calls from submerged shrubs and shoreside tangles. It is during spring courtship—from late February to May, depending on latitude—that the peeper makes itself known. Its clear, plaintive piping notes—delivered at intervals of about once a second—have amazing carrying power, and can be heard as far as a mile away on still nights. A large convocation of the tiny songsters sounds almost exactly like the ringing, jingling peal of sleigh bells, an analogy so often applied to the lilliputian frog that it is called the "sleigh bell tree toad" in some areas.

Except for those few short weeks in spring when the peepers trill their love songs lustily from every boggy and swampy spot, these frogs are furtive dwellers of the summer woodlands, and are seldom seen and almost never heard calling. Their dependence on the intact forest environment for much of their sustenance and safety during the better part of the year would seem to indicate that as the woodlands disappear under the bulldozer blade, the spring peeper will quickly follow them into the oblivion of suburbia. But the animal is surprisingly adaptable and accommodating if left with even a remnant of sheltering vegetation and semipermanent water. Though they must gather in concentrations for breeding purposes, they quickly disperse throughout the area for the remainder of the warm months and are seldom noticed or even encountered among the summer vegetation.

Small peeper gatherings often persist in low spots surrounded by subdivisions, and their sweet calls can often be heard wafting cheerily from the flooded sumps of interstate highway median strips while urban traffic thunders by on either side. One such colony I am personally familiar with exists in a low, boggy area of less than 100 square feet that lies completely encircled by a highway jug handle. Within the shadow of a nearby Sheraton hotel, the tiny frogs cheep and trill hopefully each March, accompanied by the din of passing traffic and slamming car doors at the big hotel's entrance. The eternal music of spring persists in some of the most forbidding and unpromising places, but each year there seem to be fewer singers.

Spring peepers

Chapter six

The birds

In 1985 the Food and Agricultural Organization of the United Nations conducted a survey and came up with a rather startling bird-related statistic. It seems that one of the world's most abundant vertebrates is a bird, and a very familiar one. In that year there were 8,595,760,000 domestic chickens unhappily awaiting their fate on the planet, give or take the couple of million that had just gone to the chopping block as the final count was in. The chicken census-takers determined that at the time of the count, there were 1.6 chickens for every human being alive.

At the opposite end of the scale, at least in North America, are the 76 California condors—all but five of them currently behind bars in protective custody—and the ivory-billed woodpecker, Bachman's warbler, and the Eskimo curlew. All of these are critically imperiled species that have not been reliably reported in the wild for decades, but are not yet officially declared extinct. These four species, currently lost in a sort of environmental limbo, are prime examples of those animals that are not up to the challenge of survival in the new world of humankind.

Among all of the earth's animate creations, the birds are among the most noticeable and apparent, both visually and audibly; everyone is familiar with the sight and sound of them, and the creatures have long figured prominently in the art, literature, and consciousness of humanity. The birds are the animal group most often thought of when one hears the word extinction, for their decline and disappearance is noticed much more readily than that of all other creatures—with the possible exception of the larger mammals. The birds have captivated and beguiled humankind for countless ages, and their plight today is a source of concern and distress to all who love and admire their beauty, form and action.

This book does not propose that we will soon live in a world bereft of birds. Birds will probably always be with us in one species or another, exploiting the biological niches available to them wherever the environment is capable of supporting life at all. In North America today both native birds and exotic imports are still reasonably abundant and varied. Many people who are only casually interested in nature or birds see no cause for alarm for the future of our avian fauna. On the surface, there seem plenty to go around. But it is the kind and variety of those species that are around that should be of concern to us today.

The density and variety of birdlife, of course, varies according to the habitat. There are a greater number of birds and species of birds to be found in rural areas, forests, marshes, or prairies than are found resident in urban centers. Studies conducted in London in the 1970s showed that a mere six species of birds were resident and nested in the center of the city, while at the suburban edges just a few miles away 14 breeding species were counted. Intense urbanization is hospitable to none but the hardiest of creatures, including birds. Wood warblers can still be seen during migration in the smallest of woodland fragments in New Jersey, Kentucky, and Ohio, and hawks still soar over the forested ridges of the Appalachians. There are robins and catbirds in every suburban backyard, and one can still hunt wild turkeys and bobwhite quail in Virginia. Crows, gulls, and starlings gather in millions at every landfill and airport, and there still seem to be hordes of waterfowl crowding into the refuges each autumn. In short, there appear to be all the birds the average person could hope to see, so what's the problem?

The key to the problem lies in the word *relative*. Today, we're thrilled at the sight of hundreds of waterfowl coursing above a wildlife refuge, or a group of black-bellied plovers resting on an ocean beach or saltmarsh, forgetting that these birds literally darkened the skies in their millions early in this century. Likewise, the sight of a half-dozen warblers flitting through the spring forest canopy reassures us that the eternal cycle continues, but I can remember a mere 25 years ago the May bird counts garnering 20 to 30 species of songbirds within an hour in a single woodland tract. Many of the commonly seen birds of a mere two decades ago—species like cuckoos, hummingbirds, screech owls, wood and hermit thrushes, and many of the shorebirds—are becoming harder and harder to find today. They are no longer the easy staples of the beginning birder. Seemingly suitable habitats seem empty of resident or migrant birds, as I found a couple of years ago when I attempted without success to locate screech owls in broken woodland and farm country in central and southern New Jersey. Whereas in the late 1960s a little "screechie" could be called up within minutes along these quiet rural roads, none were found in several hours of diligent whistling and peering into the wooded gloom. And the roads themselves are now paved and much busier with traffic. The screech owl is not considered a threatened species, but the question is: Why are the woodlands that harbored so many of these little avian gnomes a few short years ago empty of them today? And is this true of other regions within the bird's range as well?

Approximately 99 species of North American breeding songbirds migrate south to spend the winter in Mexico, Central America, and the Caribbean. Wintering North American songbirds average 30 percent of the total avifauna at seven sites in Mexico, 12 percent at eight locations in Central America, and from 1 to 30 percent at 28 locations in the Antilles. Twenty-nine species of warblers alone winter in southern Mexico; an additional 53 species of songbirds range as far south as South America.

The hooded warbler is one species among 16 wood warbler species that have shown sharp declines in the Smokey Mountains over the past 20 years. A study of a tract of virgin forest in West Virginia showed significant declines in populations of most migratory songbirds over the past 35 years.

The loggerhead shrike, as well as several shrike species worldwide, has undergone a dramatic decline that has biologists worried. The loggerhead, or "butcherbird," a robin-sized predator that hunts insects, small birds, and mice, and is best known for its habit of impaling prey on thorns and barbed wire, was once a common bird throughout much of North America. Today the species is listed as extinct in Maine and Pennsylvania, endangered in 11 other states and threatened in two more. No one is sure just why these birds are vanishing, but habitat loss and pesticides are high on the suspect list.

These statistics and observations show that as both the temperate breeding woodlands and the tropical wintering grounds—the rain forests—are dismantled for housing and agriculture, the birds are displaced and gradually fade away. These species simply cannot adapt to bisected and fragmented habitats. They become vulnerable to human disturbance and the invasion of hardy and aggressive forest edge interlopers like the grackle and cowbird.

For the most part, those birds that will be with us in good numbers tomorrow are those species that are the most numerous and evident in our midst today. Most, but not all, of the bird survivors are those hardy species that are resident wherever they occur or, at best, have shorter migrational shifts according to the seasons. Crows, starlings, house sparrows, mockingbirds, and cardinals are examples of these. At-risk birds are those that migrate between the forests of North America and those of the tropics—both environments under relentless assault today and rapidly shrinking. The red-eyed vireo has been called the most abundant songbird in North America, with population densities of up to five pairs per acre of forest, but this bird, too, is probably doomed, for it cannot adapt to the loss of intact, relatively unfragmented woodland. The same is true for various chickadee species and such warblers as the familiar yellowthroat and yellow-rumped. All are still fairly common today, but they require the kind of extensive, vegetated habitats that are not provided by the spreading suburban and urban infrastructure.

The negative effect of deforestation on migrant birds here in North America is only now being fully realized, but its impact on the local level has been studied for a long time. The case of Cabin John Island, a small enclave of riverine forest in the Potomac River near Washington, D.C. is typical. In 1958 an extensive area of forest on the nearby riverbank was cleared to make way for an expressway. Prior to the destruction of the forest, Cabin John Island was home to a large population of migrant songbirds that bred there, but following the removal of the forest cover nearly all resident species declined. The redstart, which had been the most common bird on the island before construction of the parkway, declined within two years from 14 pairs to about five. Other species, such as the Kentucky warbler, vanished altogether.

Larger bird species, such as the various gulls, the black and turkey vultures, the ring-necked pheasant, and the Canada goose have demonstrated an ability to exploit the prevailing habitat that few others have. While other water and game birds with more specialized feeding and habitat requirements decline, these species and a few others have thrived and expanded their ranges, often depending on the largess of human bird lovers, and even becoming nuisance or pest species in some cases.

The bottom line of bird survivorship is that the birds, more mobile and less physically tied to specific places on the land or waters than more earth-bound organisms, are better equipped to move about and effect population shifts in the face of spreading habitat degradation. Nonetheless, only those species able to exploit those new environments will ultimately survive. Whenever you watch herring and ring-billed gulls squabbling over bread handouts at the beach or garbage at the local dump, starlings usurping nest sites from other, less assertive birds, or house finches setting up housekeeping in a traffic light above a busy intersection, you are seeing first-hand the survivorship qualities of birds.

House sparrow
Passer domesticus

Worldwide, at least 1197 introductions of exotic bird species have been made over the past two centuries. Some 576—or about 48 percent of the total number of these attempts to bring birds to new and alien habitats—can be considered failures. The introduced species was either quickly eliminated by resident predators, or could not adapt to and exploit the environmental conditions it encountered. In North America 53 percent of the avian introduction attempts made since the year 1800 met with failure.

One of those bird species that beat the odds and settled down here to stay is that little feathered ruffian known as the house sparrow.

Also called the English sparrow, hoodlum (obsolete), and domestic sparrow, this chunky, aggressive little Old World weaver finch—an illegal alien if there ever was one—has become the dominant "songbird" of the cities and suburbs of North America. William Cowper, the 18th century English poet and apparently no fan of the house sparrow, wrote of the bird: "The sparrow, meanest of the feathered race, his fit companion finds in every place, with whom he filches grain that suits him best." These words suit the sparrow well, for though it is well-known and tolerated by Americans today, it is certainly not loved by them, and most people are at least vaguely aware that the bird is an alien interloper that has brought misfortune to our native birds.

House sparrow

♀

♂

Historians believe the house sparrow has been associated with mankind in the Middle East since true agriculture began around 8000 B.C. The word "sparrow" itself stems from *zippor*, the ancient Hebrew generic term for bird. Fossil remains of an extinct house sparrow, *Passer predomesticus*, carbon dated at about 400,000 years, have been found in the Judean hills near Bethlehem.

The house sparrow was one of the earliest exotic bird introductions into North America. Although scattered introductions reportedly took place as early as 1843, the first officially documented contingent arrived here in 1850. In that year eight pairs were brought from England to the Brooklyn

Institute of Arts and Sciences in New York. In the spring of the following year the birds "were liberated [in Central Park] but did not thrive," according to a report from the period.

In 1853 another importation of 50 sparrows was released near the Narrows on Lower New York Bay, and later a smaller group at Brooklyn's Greenwood Cemetery. These birds survived their first year and began to multiply; some of their progeny were subsequently "caught and taken to Cuba by Spanish monks" who apparently felt that the sunny Caribbean island just had to have a few house sparrows down among the sheltering palms.

Although reference is usually made to the Central Park release as being the birds' initial foothold on the continent, the 500-acre Greenwood Cemetery is now considered to be the true launching pad for the house sparrow's eventual conquest of North America. Part of the ensuing invasion was aided and abetted by people, for up until the end of the last century the birds were still regarded as desirable additions to the native avifauna.

Within 12 years, a number of imported sparrows were deliberately released in Portland, Maine, Rhode Island, and in Madison Square in Manhattan. By 1864 the sparrow had arrived, under its own power, at Quebec, and a few years later they were reported to be nesting in Halifax, Nova Scotia, as well as in Toronto and Montreal.

Here in the United States the sparrow's march was equally impressive. By 1886 the sparrow had invaded the 38 states and 10 territories of the nation as well as the District of Columbia. Only about 40 percent of this expanse, or 885,000 square miles, was actually occupied by sparrows, however.

Ohio was one of the few states that actually took the trouble to inventory its sparrow horde, coming up with the estimate of some 40 million sparrows occupying 20 million acres of prime habitat—namely, the state's many large and small towns and farms. Based on field observations, the tally worked out to an average of 1000 sparrows per square mile of Ohio terrain. At the time of the Ohio survey—in 1887—the nationwide population of house sparrows was estimated at about 885 million birds.

Meanwhile, by 1870 the sparrow had arrived in California, and invaded the Pacific Northwest 20 years later. Today, the bird is found from Alaska to southernmost South America in this hemisphere. It has also been introduced in many other places over the globe, including Australia and Hawaii. In New Zealand, where the sparrow was introduced early in this century, a visiting ornithologist recently witnessed one of the reasons why this little feathered rogue has been so successful in the game of life—pure street smarts. Groups of the birds hung around the main bus terminal in the city of Hamilton and were observed to deliberately activate the automatic door sensors to gain entrance into the building. The scientist reported that the sparrows scavenged crumbs from the floor of a cafe inside the terminal and quickly learned the secret of the automatic doors, opening them 16 times in the 45 minutes they were under observation. Two males from the flock of avian freeloaders would either hover directly in front of the sensor's eye or land on the box and peer into it, causing the door to open!

Although the house sparrow reached almost fabulous population densities in the last century, it is today far less abundant, and has become a more reasonable member of the North American wildlife community. The bird's gradual decline to more manageable population levels can be tied almost directly to the disappearance from the American scene of another familiar animal—the horse.

In the 19th century the motor vehicle was nonexistent and everything moved by horsepower. The millions of horses and mules required to move people and goods in the era of the first house sparrows generated many tons of undigested and wasted grain and other animal feed—the fodder that fueled the little bird's spectacular rise here. As gas-powered conveyances gained in prominence in the early years of this century, the horse faded from the scene rather rapidly. In 1919 there were 75,740 horses in New York City as compared with 108,036 just a few years prior. Denver saw a reduction of one-third of its horse population between 1907 and 1917. In rural districts, the advent of tractors ushered in a similar decline in horses and mules. In 1910 the Department of Agriculture reported 21,040,000 horses on American farms; by 1940 the number had fallen to a little more than eight million.

All of this meant far less food for the house sparrow, especially during the winter when reliable sustenance was most needed. The birds began a gradual decline that today has resulted in population densities much more in line with the carrying capacity of the sparrow's environment. This is not to say that the combative little creature is no longer a problem for native species, but that it's extreme aggression of yesteryear seems to have abated with its lesser numbers and it now coexists in its adopted home with greater harmony. The house sparrow will always be with us, but perhaps as a complement to the local avian scene rather than an overbearing pest.

Starling
Sturnus vulgaris

John R. Quinn 1991

The common—or European—starling is perhaps the archtypal example of the consummate animal survivor among the birds. Sometimes called blackbird by the uninitiated, the starling today occupies virtually every nook and cranny of the North American continent that people call home, from the towns and cities of Alaska and Newfoundland south to Florida, California, and Mexico. Living in close, ambiguous partnership with humankind, starlings are absent only in dry desert regions, extensively rural or wilderness areas, or in subarctic regions with sustained low winter temperatures. In sum, starlings can maintain themselves wherever people offer food in the form of garbage, handouts, feeder trays, and food crops, and shelter in the form of buildings, bridges, and bird houses.

Like the brown rat, the starling is a true camp follower of man, and probably the one bird species most visible and familiar to urban and suburban residents coast to coast. They are active, noisy, gregarious, and just about everywhere. They strut and waddle about the streets and byways of every city in the land, poking into trash and panhandling handouts from bird lovers, all the while screeching and squabbling and delivering their well-known call note, accurately described by ornithologist Roger Tory Peterson as "a high-pitched, descending 'feeee-uuu!'"

More frequently than street pigeons and house sparrows, starlings form enormous flocks that range widely over entire regions in search of foraging grounds, returning en masse at sunset to great roosts on buildings, bridges, and other structures. One of the largest, oldest, and perhaps most infamous starling roosts is the 125th Street viaduct on New York City's Upper West Side. Upwards of 135,000 starlings descend upon this structure each evening from all points of the compass, a fair percentage of them streaming in across the Hudson River from neighboring New Jersey.

Unlike street pigeons, the starling can sustain itself quite well without man's direct aid. The garbage banquet of dumps and city streets, as well as natural food items such as insects and crops fill the starling's undemanding menu quite nicely.

Although the starling devours great volumes of insects, harmful and otherwise, and was considered a major factor in the ultimate control of the introduced Japanese beetle, the bird is today regarded as more a pest than a pal of humankind. Its sheer numbers coupled with its bold nature have combined to make *Sturnus vulgaris* a real threat to native bird species wherever the starling occurs in any significant numbers. Starlings also raid food crops and eat grain when it is available, and they present at least a moderate aircraft-bird collision threat near urban airports. But for the most part their chief urban offenses are the screeching din of the massed, roosting hordes, and the matter of the birds' soiling anything and everything they perch on or fly over. Buildings regularly utilized as roosts might bear the weight of many tons of odorous guano, and the strongly acidic droppings also cause extensive damage to paint and corrosion of gutters and flashing.

The starling originally hailed from western and central Europe, where it was and still is a respectable member of its environmental community. Back home the bird had the usual, and natural, complement of adversaries to contend with, all of whom conspired to maintain the bird's population at levels in harmony with the carrying capacity of the environment. Within historic times it is doubtful that there were more than 100 million starlings occupying the original range. Since the mid-1800s the world population of starlings, like that of people, has soared to many billions, and the estimated North American number alone exceeds 150 million birds. It's probably a good deal higher than this conservative figure, as starling counts in 723 major blackbird roosts in the Lower Mississippi Valley exceeded 99 million birds. The total counts for these roosts, made in 1974–75, were some 438 million birds, most of them redwings and grackles.

Starlings have long had the reputation of being serious agricultural pests. The ancient Greeks regarded them as determined crop destroyers, and employed guards armed with slings to keep them at bay. By the 18th century, the accelerating growth of European urban centers attracted starlings from far and wide over the hinterland, signaling the true beginnings of the bird's symbiosis with man as we know it today. By the beginning of this century the starling had forsaken the countryside and become a true avian city slicker, a role it continues to occupy today.

The first serious attempt to introduce the starling to the New World came in 1850, just about the time the infamous house sparrow was being loosed upon the land. The effort was as unheralded as it was unsuccessful. Then in 1890 one Eugene Schiefflin, taken with Shakespeare's reference to the bird in *Henry the Fourth*, determined that he could not endure life without a few starlings in the neighborhood, nor should anyone else have to. In the spring of that year Shiefflin imported and released 60 starlings in New York City's Central Park. The birds of this first introduction quickly dispersed into the shrubbery and vanished, Schiefflin thought, for good. A year later he liberated another 40. These starlings made themselves right at home, and later that year the first active nest was found—on a window sill of the nearby American Museum of Natural History.

From there, the starling began its inexorable march across the continent in an all-too-familiar tale of wildlife introductions gone bad. Within 10 years the starling had invaded all of southern New England and most of New Jersey. By 1916 they had colonized the East Coast from Maine to the Virginia Capes and had flown westward to Ohio. When 1929 arrived, so had the starlings—in Texas and along the entire northern shoreline of the Gulf of Mexico. California fell to the winged invaders early in the 1930s.

The starling owes its amazing success to its adaptability and determination in three main areas: it is a highly social and organized bird; it is well able to exploit virtually all sectors of the habitat for food; and it can produce two and sometimes three broods each season, the four or five young leaving the nest within three weeks of hatching. The bird's pugnacious socializing is well known to most human city dwellers, as is its uncanny ability to locate a food source wherever it might be sequestered. The sight of a gang of starlings gathered around plastic garbage bags set at the curbside while a couple of their number open holes in the tough plastic to gain access to the odorous goodies within is a familiar one to urban passersby.

It is in the area of the starling's relentless fecundity and admirable child rearing practices that the bird really shines. A 1962 U. S. Department of Agriculture study found that a brood under continual observation was fed by the adults on average once every 6.1 minutes during the rearing period. Assuming that the young are fed, conservatively, 12 hours each day, roughly 118 feedings per day are logged by the hardworking caregivers. As with the house sparrow, the starling's numbers reached their apogee within about 40 years of its landing on these shores, and the population today is in fact lower than it was in the 1920s. But unlike the sparrow, the starling's leveling off in population appears to be a natural adaptation to the carrying capacity of the birds' habitats—unlimited and varied though they might seem to be. The birds are much more numerous near large cities and the landfills that receive the megatons of refuse disgorged daily by their human residents. They are less obvious in small towns and in agricultural areas than they once were.

Regardless of the starling's fortunes at any one time, it's unlikely that we'll run out of the birds in North America. Some 30 years ago ornithologist Robert Lemmon thought that there were then about 100 billion individual birds of all species alive on earth, with between 12 and 15 billion inhabiting North America. In the environmentally diminished world of the coming century, that figure might still hold true, but it's a pretty safe bet that a fair percentage of them will be starlings.

House finch
Carpodacus mexicanus

Also known as the crimson-fronted finch, red-headed linnet, burion, and redhead, this adaptable bird is a western species that has recently made its appearance "back East," where it has become a prominent, if not welcomed part of the urban bird fauna.

House finch

The house finch occurs naturally throughout the West from southern Canada to southern Mexico and east to about Nebraska. Through natural migration and deliberate introduction by man—several dozen caged birds were released on Long Island, New York in the 1940s—this bird is now widely distributed throughout the East from New England to northern Florida and northern Louisiana. The feisty finch has tended to dominate the house sparrow wherever the two species occur together, so that in some areas the sparrow has actually declined in numbers in the face of competition with the finch.

The house finch, when compared with the introduced sparrow, is a rather attractive bird. Slightly slimmer of form, the male is brown with a rose-red head and breast, these colors becoming more intense during the breeding season. The finch also has something of a song, delivering a pleasant chippering warble that is positively musical in contrast to the house sparrow's monotonous and tuneless chirps and cheeps. Although the finch is a newcomer in the East, it is a genuine North American native, a pedigree that appeals to most people and leads many to encourage it to nest around the yard, something few homeowners are inclined to do for the house sparrow.

Pretty and pert though they might be, house finches are aggressive and inventive nest builders and a pair will set up housekeeping in any cavity or container that can be defended against takeover by other house finches. Hanging porch plants, light fixtures and drains have all been appropriated by the house finch, and the bold little creature will immediately accept small baskets or pots hung for them on porches and patios.

In general, the house finch seems less intimidated by the close proximity of people than that arch survivor among birds, the house sparrow, which usually nests well out of reach of people and their pet cats.

Common and fish crows
Corvus brachyrhynchos and *ossifragus*

Also called the American and carrion crow, this conspicuous bird has extended its range and influence far beyond the rural environments with which it was once associated. Occurring from British Columbia and Newfoundland south to Florida, the Gulf coast, and northern Mexico, the common crow is certainly no longer restricted to farmyards and cornfields, but can be found in the parks, open areas, and along roadsides at the very centers of our largest cities. The suburbs from coast to coast are well stocked with the common, western, and fish crows and winter roosts of more than a half-million birds are not at all uncommon.

The fish crow, a slightly smaller bird with a higher-pitched, staccato *ca-a* call as opposed to the common crow's unmistakable *caw*, is a coastal species occurring from New England to the Gulf of Mexico. This crow has been expanding its range inland in recent years, following the courses of larger rivers.

Both crows are all-black birds, but they are not true blackbirds—a group containing the grackles, meadowlarks, orioles, and the cowbird. Crows are more closely related to the jays and magpies.

In discussing the survivorship of the crow, a study in contrasts is called for. All of the larger hawk species are sharp-eyed, powerful predators well able to take care of themselves against their natural enemies—including crows, which any accipiter, or bird hawk, can dispatch without trouble. The goshawk, for example, has few peers when it comes to speed and maneuverability in the air, even through dense forest, and its vision allows it to spot prey or potential trouble at the slightest movement. Yet the goshawk, like most raptors, is currently declining, while crows seem to be more abundant than ever. Why is this the case?

 A closer look at both birds will reveal at least part of the answer. The hawk, for all its speed, power, and visual acuity, is not an overly bright creature; it operates almost entirely on instinct. As long as its habitat remains intact, and the threats it faces fall within the parameters it is genetically designed to cope with, it can avoid environmental and physical danger. The crow, on the other hand, lacks the killing talons and hooked bill of the hawk, but it is far better able to adapt to rapidly changing factors in its environment, and to "learn" from experience. No matter how many goshawks have been shot by hunters, this deep-wilderness species seems unable to recognize a man with a gun as a distinct threat, and many are still killed by firearms. Crows are completely indifferent to people without guns or carrying shopping bags, but they quickly vanish or keep their distance at the sight of a firearm in hand, or in rural areas, where they seem to know that people often carry firearms.

Crows have been implicated as serious predators on the young of songbirds and game birds, and as crop destroyers, but as they also consume great volumes of insect pests the debt is usually paid in full, except where the birds are inordinately numerous. The birds are ardent scavengers, and virtually every highway and byway in the land will have its crow "roadkill patrol" that quickly disposes of the horrendous toll of animal life killed annually on the roads of North America.

Crows have long been popular as pets, and adjust well to human company if taken young enough. There are drawbacks to keeping a crow around the house, however, for a pet crow is rather like a feathered monkey—full of insatiable curiosity and a definite penchant for mischief. Although they are regarded as destructive birds in many states and thus unprotected by law, their possession as pets by private individuals is generally discouraged, or prohibited outright.

Common grackle
Quiscalus quiscula

Also known as the crow-blackbird, purple or bronzed grackle (races), blackbird, and keel-tailed grackle, this sleek and elegant bird is well known to every urban and suburban resident east of the Rocky Mountains. It is one of the commoner lawn birds of the east, gleaning the insect harvest along with starlings, crows, and house sparrows.

The grackle's wide range includes the length and breadth of central Canada south to Texas and Florida. The average person knows this bird simply as "blackbird" and usually distinguishes it from the smaller and stockier starling.

At one time the two races, or varieties, of this species—the so-called purple and bronzed grackles—were considered to be separate and distinct species. Both varieties are vividly iridescent with shades of purple, blue, bronze, and green; wherever they occur together various hybrids are produced. One of these hybrids, which combined the bronze and purple hues in the form of indistinct barring, was once called "Ridgeway's grackle," but the two races have since been combined into the common grackle. The taxonomy of the species might yet change again, but at least the birds know who they are!

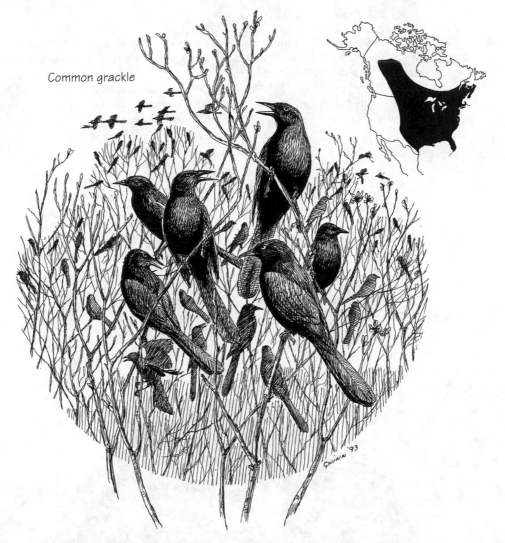

Common grackle

The grackle is an opportunistic feeder, consuming everything edible from seeds, nuts, and insects to carrion and the eggs and young of other birds. I have seen this bird attack and kill house sparrows, swooping in on a flock feeding on a lawn, nailing a single bird and flying off with the limp victim, just like a miniature hawk. Certainly not the behavior one would expect from a neighborhood "songbird"!

Grackles have increased vastly in number since Europeans came to this continent. The great changes that have occurred in the North American environment have, for the most part, benefited them. Although the birds are subjected to large-scale control programs from time to time, these have not diminished their populations to any appreciable degree; the grackle is an extremely prolific creature. Winter roost counts have shown that this bird is second in abundance only to the red-winged blackbird. There is no way to estimate accurately the total North American grackle population, but it is doubtless in excess of 300 million birds.

Red-winged blackbird
Agelaius phoeniceus

Also known as the swamp blackbird and red-shouldered blackbird, the red-wing is one of North America's most abundant native birds. A population survey of the species undertaken in 1983 by the

U.S. Fish and Wildlife Service revealed that the species had a population of 25 to 26 million birds. Today that estimate has been increased to about 30 million, though it is very probably a good deal higher. For example, in 118 eastern U.S. blackbird roosts studied in the 1970s, it was estimated that 190 million of the birds were red-wings, 110 million were common grackles, followed by 99 million starlings, 91 million brown headed cowbirds, and 10 million Brewer's blackbirds. Three hundred sixty-five roosts in California and Texas contained roughly 139 million birds, the majority of them red-wings.

Red-winged blackbird

Blackbird specialist Gordon Orians in his book, *Blackbirds of the Americas*, notes that open-country blackbirds like the red-wing are "far more abundant today than at any time in the past" and have benefited from man's alteration of the habitat. On the other hand, he observes, island and tropical forest blackbird species are as vulnerable to habitat destruction as any other bird species and many are in a precarious state today. "Life will be a bit poorer if fewer than ninety-four [blackbird] species survive with us into the twenty-first century," Orians concludes.

But for now, things look rosy for the red-wing. The bird has one of the most expansive ranges among resident songbirds, occurring from south-central Alaska and subarctic Canada south to Central America and Cuba. The bird is resident year-round in all parts of its range except for the most northerly parts, where it effects regional shifts with the onset of cold weather.

Although the red-wing prefers nesting areas that are near water, it is not inflexible in this respect, and its willingness to accept varied habitats has made it an avian survivor. The species will form loose nesting colonies in freshwater marshes and bog edges, but many are observed in dry fields, blackberry tangles, suburban parks with areas of thick vegetation, and even on highway median strips where there is enough cover. The nests are usually some distance apart as the pugnacious males hold and defend sizeable territories.

Cowbirds
Molothrus ater and aeneus

The cowbird, also known as cow bunting, cow blackbird, buffalo bird, or cow-pen bird, was a close companion of the once-abundant buffalo, or bison, but has taken the path to survivorship rather than near-extinction. Always abundant, the cowbird population has swelled enormously with the advent of urbanization. Today it is one of the most numerous of blackbirds, gathering in winter roosts numbering in the many millions. The brown-headed cowbird, through no fault of its own, is considered to be one of the real "bad guys" among birds due to its less than responsible method of child rearing. No cowbird chick has ever known its biological parents, for the species is entirely parasitic upon other birds and leaves its parental duties to them by laying its eggs in their nests while the rightful owners are away. Through this behavior, the species has been implicated as a major factor in the disappearance of more desirable songbirds. The problem has two root causes: the increasing abundance of the cowbird, which has found cattle ranches, farms, and suburbs to its liking; and the fragmentation of woodlands by development and road building, which permits the parasites easier access to the nests of deep-forest species.

This parasite among birds lays between three and six eggs, all of them in the nests of unwilling host species. Finches, warblers, and vireos are those birds most often victimized, though thrushes and other blackbirds might also find themselves with a cowbird chick on their hands, or wings.

Cowbirds form no lasting pair bonds; a single female will be escorted by several males, who station themselves in the nearby treetops while she searches out suitable nests in which to deposit her eggs. Once a nest is found, the bird waits until the rightful owner leaves, quickly slips in and removes an egg, leaving one of her own in its place. Some hosts spot the alien egg and either eject it or rebuild the nest over it, but most host parents are fooled into incubating it and rearing the huge cowbird chick that soon emerges. In most cases, the cowbird chick hatches first and either crowds the rightful young out of the nest or causes them to starve as it demands nearly all of the parent birds' time and attention.

During the breeding season cowbirds disperse far and wide in search of nests to pilfer and victimize, but in the winter they assemble in vast flocks with other blackbirds in farm fields, landfills, and marshes. As the fragmentation of forests continues in North America over the coming decades, the cowbird can only benefit from these changes in the landscape. Along with other open-country blackbirds it will probably continue to multiply, posing an increasing threat to whatever small songbirds still manage to hang on.

In spite of its bad attributes, the brown-headed cowbird is not an unattractive creature. A bit smaller than a robin and the smallest North American blackbird, the male cowbird is a glossy greenish-black with a dark, chestnut-brown head; his mate is a uniform, soft grayish-brown. It has a short, stubby, rather finchlike bill, unlike the longer more pointed bills of most other blackbirds. It is generally nonaggressive around bird feeders, and outside of the breeding season gets along well enough with all other species.

The bronzed cowbird (M. aeneus) is the western and tropical version of the brown-headed cowbird, occurring from the extreme southwestern United States in Arizona and Texas south to Panama. Like the brown-headed, this cowbird is parasitic, surreptitiously laying its eggs in the nests of other blackbirds, orioles, and other species of similar size. It is highly social, consorting with other blackbirds and forming huge roosts in suburban and city parks throughout its range.

Mourning and ground doves
Macroura carolinensis and Columbina passerina

Also known as dove, turtle dove, and Carolina dove, the mourning dove is very nearly as abundant and visible as the introduced rock dove, or street pigeon. Protected as a songbird in most northern states and hunted as a game bird throughout much of the South, this slim, sleek, swift-flying bird has adapted well to suburban and urban environments, and readily nests in close proximity to people and their dwellings.

Mourning dove

The mourning dove occurs from extreme southeastern Alaska and New Brunswick south to Panama and the West Indies. Throughout much of its extensive range—all but the very coldest portions—it is resident virtually year-round and is a regular visitor to backyard bird feeders. The species is one of the relatively few that has directly benefited from the removal of much of North America's original forest cover and the burning off of the vast, midcontinent grasslands. The disturbed environments of today seem perfectly suited to this adaptable bird, which has made itself at home in waste habitats that could not support many other species. Wherever cover composed of hardy, seed-bearing weed plants appears on degraded landscapes, there too will appear the mourning dove, and often in considerable numbers. Landfills, fallow fields, airport verges, weed-grown construction sites, and many other such disturbed and damaged habitats are perfectly suitable dove habitats as long as the weed cover remains.

Reaching a length of about 13 inches, this dove is a light gray above, pale buffy brown below. The wings are darker than the body and are adorned with a series of black spots along the inside edges. It has a pale blue eye ring and a patch of violet-blue iridescence on the neck. Its name stems from the mournful, owl-like cooing call of the male, usually delivered from a prominent perch and often the prelude to an elaborate courtship flight and display.

Small groups of these doves usually feed quietly on the ground among weed cover and are not noticed until they explode noisily into flight and dart erratically but swiftly off.

The mourning dove nests two to four times a year, depending on latitude, and the two young, or squabs, are fed a liquid secreted in the crop, called "pigeon milk." The species' impressive fecundity helps assure its continued abundance in the face of considerable hunting pressure in southern states. The tiny ground dove (*Columbina passerina*) is a familiar suburban and roadside bird of the South and Southwest. It is also widespread throughout Latin America. The ground dove reaches a length of about six inches and is a plain, soft brown, much like the mourning dove. The male has a blue crown and shows much iridescence on the neck feathers, The body feathers are dark-edged and give the bird a scaled look.

Inca dove

Inca

QUINN '93

Ground dove

The equally pint-sized Inca dove (*Scardafella inca*), a paler version of the ground dove, has spread northward from Central America into the extreme southwestern United States, where it has become an increasingly familiar bird of farms, cities, and suburbs. The reliable presence of water in these urban oases in the desert scrub has greatly facilitated the spread of these engaging little pigeons. Like the ground dove, the Inca dove is an extroverted, fast-flying bird that shows little fear of man.

Rock dove
Columba livia

Also known as the street pigeon, city pigeon, and domestic pigeon, this ubiquitous bird needs no introduction. Everyone is familiar with the city pigeon, and the sounds and mess it generates. The rock dove is literally cosmopolitan in distribution, being found nearly worldwide in all climes and on most terrains. The bird has been transported by man to every continent except Antarctica. In North America it is found throughout the continental landmass with the exception of the far northern boreal forests and tundra regions. In short, everywhere that mankind calls home, the rock dove does as well.

This dove is the ancestor of all domestic pigeon breeds, and even in cities the color and pattern of feral birds can vary widely, from pure white through piebald browns, reds, blues, and grays, to the ancestral color: blue-gray with with a white rump, a black terminal band on the tail, and two broad bars on the wings. There is an area of rainbowed iridescence on the neck, especially noticeable in the courting male.

In the wild the rock dove nests on the high sea cliffs of the European Atlantic and Mediterranean. Wary and swift of wing, it has few enemies in the wild state except the fleeter hawks and falcons that easily pluck it from the skies in flight. Although peregrine falcons take an occasional pigeon in and near North American cities, these birds of prey are nowhere near numerous enough to make a dent in the local pigeon population, and it is this absence of natural enemies that has made the bird's rapid spread here possible—along with the dedicated legions of urban pigeon feeders (few rural folk hold any affection for the birds).

Although the rock dove can and does survive in some places as a truly wild bird independent of man, this is unusual; the great majority are quite reliant upon deliberate feedings by pigeon lovers or on the leavings of humanity found in landfills, street trash, and dumpsters. The bird is an industrious gleaner and grazer, however, and flocks are commonly observed inspecting the most spartan terrain—from ocean beaches to mall parking lots—for edibles.

Rock doves can nest two or three times a year, setting up housekeeping in the middle of winter if it is a mild one. The usual clutch is two plain white eggs deposited in a crude nest of sticks and debris placed on cliffs, buildings, bridges, and any other suitable constructions. Pigeons are highly social birds and they both nest and flock close together. A gang of ordinary street pigeons in flight presents a spectacular sight, no matter how you might otherwise view the birds, for they engage in aerial acrobatics in perfect unison like so many sandpipers, and exhibit a joyful exuberance in the experience of flying.

The pigeon is one of the swiftest and strongest of fliers among the more familiar birds; only the peregrine falcon can routinely outfly and catch a fleeing rock dove moving at top speed. Racing pigeons have been clocked at 55 miles per hour under normal conditions, while a few individuals, assisted by a good, stiff tail-breeze, have reached a reported 90 miles per hour over the short run. One bird was reliably clocked at an average speed of 72 miles per hour over a 200-mile course. Racing rock doves have been known to cover nearly 600 miles distance over 12 hours.

Despite periodic and generally ineffective control and eradication efforts, pigeon numbers are as high as ever, and the birds are likely to be with us for a long time to come. The two crucial factors are an abundance of nesting structures and a reliable food supply—two commodities not likely to disappear from our spreading cities in the near future!

Gulls
Larus species

Generically called "sea gulls" by most people, the gulls as a group are among the most adaptable of the larger avian creatures. Not all gull species, however, can be considered survivors or are even able to hold their own in a rapidly changing world.

North American gulls range in size from the big and powerful glaucous and black-backed gulls, which can be 30 inches long, to the dainty, ternlike Bonaparte's gull, adult at about 12 inches. In between are the California, herring, ring-billed, and laughing gulls—the species most familiar to the majority of Americans and Canadians and the more visible "sea gulls" of the coasts and larger lakes.

The herring (L. argentatus) and ring-billed (L. delawarensis) gulls are what could be considered the most typical gulls, and have become more closely associated with mankind over the past century. Slaughtered without mercy in the early years of the 20th century for the millinery trade, they now enjoy full protection and a vastly increased and reliable food supply near urban centers. These gulls, as well as the smaller laughing gull, have thus become much more numerous than they ever were in the past. The herring gull, originally a bird of the Canadian arctic, the Maritime Provinces, and northern New England, has been steadily extending its range to the south as urbanization and its attendant refuse and landfills offer boundless new food supplies to the voracious birds. This gull, once unknown as a breeder south of Cape Cod, has been reported nesting as far south as coastal Virginia and North Carolina.

Great black-backed gull

Herring gull

Gulls, like the imperiled terns, are ground nesters that gather in fairly large breeding aggregations, but unlike the latter, which are restricted to specific shoreside habitats, the gulls will nest on any flat, isolated terrain where they sense they will be free of persecution by people or predators. This can include abandoned landfills, dredge spoil banks, airports, inactive coastal construction sites, and even temporary sandbars and marshy islets. Whereas the endangered least terns and piping plovers require specific zones on oceanfront beaches for nesting—places also highly popular with growing armies of human sun worshippers—and are disappearing as a result of disturbance, the adaptable gulls will simply move on and find other, less peopled areas for their reproductive activities.

Wherever they are overly numerous gulls pose problems on several fronts. Considering the huge volume of aircraft flights daily entering and leaving the nation's major airports, incidences of gull-aircraft collisions are very few in number, although this threat gets the most media play. During the 1991–92 breeding season, sharpshooters were employed by the Port of Authority of New York and New Jersey to destroy a percentage of the swarms of laughing gulls congregating around the sprawling JFK International Airport. Between 10,000 and 15,000 of the birds were reportedly killed during the control campaign.

The larger gulls are also accused of destroying the eggs and young of the smaller terns and breaking up their already threatened coastal colonies. Large herring and black-backed gulls will attack and eat any bird or animal it can catch and overpower and they can wreak havoc among nesting waterfowl. I once witnessed an attack by black-backed gulls on a brood of nine mallard ducklings in which, despite the frantic efforts of the hen to protect them, all of the young were plucked from the water and carried off by the big gulls.

Ring-billed gull

Laughing gull

In spite of their environmental drawbacks, gulls perform a valuable scavenging service in the human-dominated environment and more of the birds will be found in the vicinities of large, active landfills than nearly anywhere else. Although many birds, especially the less experienced juveniles, perish each year from botulism and contamination by toxic wastes, the vast dumps remain the primary food source for the majority of urban gulls today.

The California gull (*L. californicus*) is similar to the herring gull but is smaller and has a darker back, or mantle. Its eye ring is red and it has green rather than yellow legs. This aggressive bird is found throughout the northern prairie states and provinces, westward to northeastern California. It forms huge nesting colonies in shallow, inland lakes, often in consort with the smaller ring-billed gull. This species attained everlasting fame as the savior of the Mormon's first wheat crop in what is today Utah. It arrived in time to destroy the locust hordes poised to devour the crop, and a monument commemorating the event stands in Salt Lake City.

Cattle egret
Bubulcus ibis

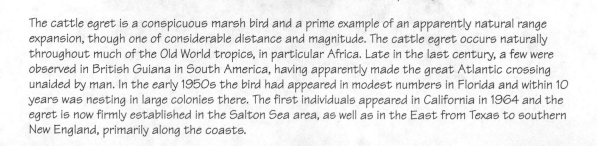

135

The cattle egret is a conspicuous marsh bird and a prime example of an apparently natural range expansion, though one of considerable distance and magnitude. The cattle egret occurs naturally throughout much of the Old World tropics, in particular Africa. Late in the last century, a few were observed in British Guiana in South America, having apparently made the great Atlantic crossing unaided by man. In the early 1950s the bird had appeared in modest numbers in Florida and within 10 years was nesting in large colonies there. The first individuals appeared in California in 1964 and the egret is now firmly established in the Salton Sea area, as well as in the East from Texas to southern New England, primarily along the coasts.

In North America cattle egrets are birds of the wide salt marshes, though they occupy pretty much the same niche as in Africa, where they are followers of cattle. Here too they are often seen riding on the backs of grazing cattle and feasting on the insects flushed by the wandering animals' hooves.

The cattle egret is most commonly observed in the broad coastal marshes that still fringe the continent's eastern shore, and they are especially common in the larger wildlife refuges of the East Coast. They constitute a rather small percentage of the nesters in heron rookeries, and often associate with the more common species such as the snowy and common egrets and the black and yellow-crowned night herons.

The cattle egret has not, so far, created a problem for native birds, though it has been accused of usurping the best nesting sites in heron rookeries and attacking and devouring the young of water-fowl and game birds.

Mockingbird
Mimus polyglottos

Also known as the mocker, mimic thrush, and mocking thrush, the brash and engaging mockingbird has become a familiar sight in the northern states, where it was virtually unknown thirty years ago. Formerly thought of as a bird of the Deep South, the mocker has extended its range north so that it is now found from southern Oregon and Utah east to Newfoundland and south to Mexico and the Caribbean. It is resident year-round in all but the coldest northern and northwestern parts of the range.

The mockingbird is robin-sized, but slimmer and more streamlined. It is gray above, whitish below, and has bright white patches on the wings that flash conspicuously when the bird flies or engages in the curious habit of "wing raising" while feeding on the ground. The species is highly territorial; a male will not tolerate the presence of another male, or a female out of the breeding season, anywhere within his strictly defined and enforced turf boundaries.

The fabled song of the mockingbird, often delivered throughout warm, moonlit nights in spring, can be next to impossible to describe. It might consist of imitations of as many as 30 other bird songs as well as various other natural and manmade noises. In general, though, it can best be described as a long series of musical, bubbly or rasping notes and phrases, delivered several times in succession, and usually from a prominent perch near the center of the bird's territory. Its alarm or aggression call is a sharp chuck! or chack!

Originally a bird of broken, shrubby woodlands, canyons, and marsh edges, the mocker has taken readily to the urban environment and is one of the commonest lawn and garden birds today. It requires broad open areas for feeding and dense shrubbery for nesting, and both of these commodities are amply supplied by the older, more established suburban habitats. It is also common in large cemeteries and rural farming country, as well as in any open areas with scattered thickets and copses. This alert and aggressive bird can be quite secretive and crafty when setting up housekeeping. It places its bulky nest of weed stems, sticks, and assorted other debris in a dense, low bush or conifer, often quite close to human habitation. The nest's location is seldom suspected or discovered until the plant's leaves fall in autumn.

Mockingbirds appear to be maintaining high numbers, even in areas under intensive development pressure. As long as enough ornamental and natural vegetation is present for both cover and nesting, it is doubtful that the inventive and irrepressible mocker will suffer any real decline in the foreseeable future.

Cardinal
Richmmondea cardinalis

The cardinal, also called redbird, is primarily an eastern bird, though an isolated population does reach the Colorado River Valley and parts of Arizona and Mexico in the west. The species has been successfully introduced into the Hawaiian Islands and in the environs of the Los Angeles megalopolis.

Throughout its range the cardinal is a year-round resident, establishing territories in wooded swamps, woodland edges, and in the parks and gardens of cities and suburbs. Preferring the close proximity of people, the bird can be heavily dependent on the generosity of backyard bird feeders for survival during especially hard winters. The bright red of the male, accentuated by his high crest, black face, and stout red bill, offers a vivid and welcome counterpoint—a splash of the tropics—to winter's somber grays and whites. The female is buffy-brown, tinged with red on her crest and wings and tail. The bird is so-called because of its fancied resemblance to the red robes and high, pointed mitre worn by Roman Catholic cardinals.

The cardinal is primarily a ground feeder, and thus does best where the winters are less severe and the snow cover less deep and persistent. Like the mockingbird, it favors habitats that feature larger open areas for feeding and dense shrubbery and thickets for nesting, these requirements often sup-

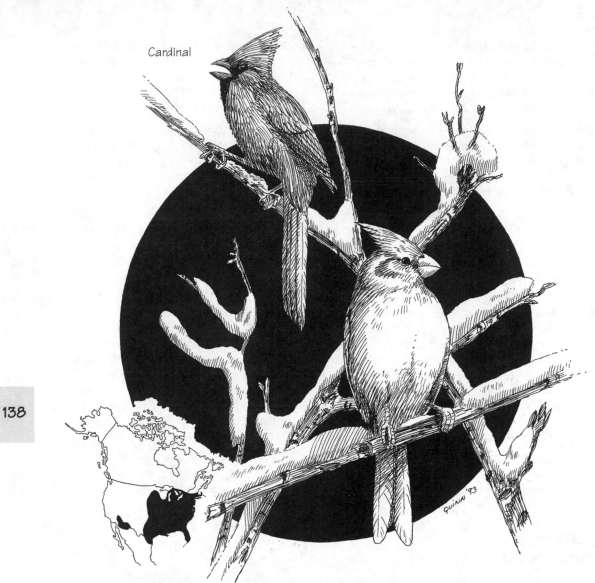

Cardinal

plied by the typical suburban yard and garden. Where the cardinal finds things to its liking it will set up its territory and vigorously defend it against all other cardinals. A spring male will instantly and pointedly respond to a whistled imitation of its call, which is a rich, far-carrying *what-cheer, cheer, cheer, purty-purty-purty!* delivered from a high perch in a tree or on a building.

The cardinal is in reality a large finch, and seeds form a major part of its diet, though fruits, berries, and insects are eaten during the warmer months.

Catbird
Dumetella carolinensis

Now officially known as the gray catbird (there is an all-black tropical species found in southern Mexico), this bird is also known as chicken bird, black-capped thrush, and slate-colored mockingbird.

The gray catbird has a very broad range, though nowhere but in the East can it be considered an abundant bird. In the West the catbird occurs from the Puget Sound area of British Columbia south to Washington State and Arizona and thence east to Nova Scotia and Florida. It is a year-round resident in southerly areas without killing frost, though individuals might migrate as far south as Panama in winter.

Catbird

Tufted titmouse

Towhee

The natural habitat of the catbird is dense, brushy, deciduous woodlands, forest edges, and stream borders, but with the encroachment of suburbia into increasing areas of these habitats, the bird has adapted well to the backyard environment as long as it is not overly trimmed and manicured. Few catbirds will be found in the biological desert that is a brand-new housing development, but if the new homeowners allow their yards to assume a reasonably natural look over the first few years, it won't be long before a catbird pair will take up residence.

The catbird is readily identifiable by its form and color; it is a slim, long-tailed relative of the familiar mockingbird, but unlike the conspicuously patterned mocker it is solid gray with a black cap and reddish or rusty undertail coverts. This bird is quite talkative, giving vent to a variety of squealing and mewing calls that have earned it its common name.

The secret to the catbird's suburban survivorship lies in its ability to be assertive and inconspicuous at the same time. The bird responds boldly to any threat and makes plenty of noise poking about on the ground beneath the backyard bushes, but it conceals its nest with consummate skill, even in rather small garden thickets, and skulks about like a feathered mouse when anywhere near it. Unlike the mockingbird, the catbird never announces its presence far and wide from an elevated perch but warbles its pleasing, musical song from the depths of a thicket or vine tangle. Because it is one of the mimic thrushes, it is a fair imitator of the songs of other birds.

Towhee
Pipilo erythropthalmus

The towhee is also called rufous-sided towhee and chewink, after one of its characteristic calls. This attractive and active bird, slightly smaller and slimmer than a robin, occurs from Maine, southern Ontario and Manitoba south to Florida and the Gulf Coast. It is a familiar sight in cutover pine and oak woodlands, waste areas, brushy fields, and older suburban parks and residential areas, where its noisy scratchings among the fallen leaves on the ground readily give its presence away.

The male towhee is a singularly attractive bird, with black head and upperparts, clear white underside and bright reddish or rufous patches on the flanks. The female shows warm reddish brown where the male is black. The white wing spots and bars flash conspicuously when the bird flies. Both sexes have bright red eyes that are quite noticeable at close range.

The towhee can be found in chaparral in the west and in pine barrens in country and suburban parks in the east; it is absent only on the broad, treeless prairies of the Midwest. Although northwestern birds and those of the mountainous areas move to the lowlands of the south-central states and Mexico during the winter, the majority of towhees are year-round residents wherever they are found. The bird conceals its loose cup of a nest well in a dense thicket or evergreen tree, often close to the ground.

The towhee's songs and calls are highly variable across its wide range, with western birds sounding very different than their eastern counterparts. The typical call is a few low introductory notes followed by a musical trill usually described as sounding like *drink your teeee!* Western birds sound a buzzy trill and an inquisitive mewing note.

Song sparrow
Melospiza melodius

Also called hedge sparrow, ground sparrow, and swamp finch, the song sparrow is one of the most familiar native songbirds. The species is one of the most widely distributed and geographically diverse of North American birds as well, with distinctly different field characteristics depending on the locale. This sparrow is found from the Aleutian Islands east across central Canada to Newfoundland, and south to central Mexico, Nebraska, and North Carolina. There is a general withdrawal southward in severe winters.

The thirty-four recognized subspecies of this bird might be so unlike each other that they seem completely unrelated. Song sparrows inhabiting the cold, humid Aleutian Islands are dark brown—almost black—and have heavy bills, while those living in desert scrub in the Colorado River Valley are a pale sandy color and have short, stubby bills. In between is the familiar song sparrow of the east—a heavily streaked, brownish little bird with a prominent chest spot and a powerful, irrepressible voice. Its cheery, familiar song is described as consisting of three short introductory notes followed by a varied trill. It has been spelled out phonetically as *Madge, Madge, Madge, put-on-your-teakettle-kettle-kettle!*

Song sparrow

The song sparrow's natural habitat consists of forest edges and clearings, open freshwater marshes and thickets, and older suburban areas. It artfully and carefully conceals its cuplike nest in low shrubs and dense tangles; up to three broods might be reared in one nesting season. This bird freely nests in gardens and city parks and industriously scratches about on the ground in shrubbery and on open lawns in search of insect fare. As long as patches of unkempt, brushy woodlots and lushly vegetated suburban yards persist the song sparrow should be able to hold its own, for at least the foreseeable future. The bird's uncritical habitat needs, as well as its ability to adapt to the close presence of people and their domestic animals will serve to make this little songster one of the fortunate avian survivors of tomorrow.

American robin
Turdus migratorius

Robin

Also called robin redbreast, redbreast, and northern robin, this beautiful and beloved bird occurs throughout most of North America, from the arctic slope and interior Alaska east to Labrador and south to California and Georgia; it has expanded its breeding range to the south over the past 25 years with the development and growth of towns and cities. The robin generally withdraws from the colder northern portions of the continent during the winter, with most birds spending the winter in the Gulf States feeding on the winter berry crop.

The robin is so familiar as to scarcely require any description at all. Gray above, rusty red or orange below, its robust form and habit of making short runs over shady lawns, punctuated by periods of intent "listening" with head cocked for earthworm prey, are a time-honored part of the summer scene. The sexes are pretty much alike in appearance in the robin, though the male has a black head and tail while those of the female are paler.

The rich, warbling *Cheer-up cheerily, cheer-up cheerily* is the archetypal song of spring and is known and loved by every winter-weary human within the bird's range. The robin also expresses alarm with a loud, ringing *put-put-put* and has a vibrant call note that can best be described as *weeep!*

The robin was originally a bird of the unbroken forests, but because it prefers woodland edges and meadows interspersed with forest it has become much more abundant since the advent of Europeans on this continent. The species is still found in some abundance in forested areas, though these "wilderness" birds are far less trusting than those that live cheek by jowl with mankind.

The robin builds a substantial nest of grass and plant fibers reinforced with mud and twigs, placed in trees or almost anywhere on buildings and other structures that afford protection from the elements and predators. Its location can vary according to the time of year; early nesters subjected to cold spells usually nest in dense conifers closer to the ground, while pairs nesting in hotter weather place the nest high in trees, a location that affords shade and cooling breezes to the incubating hen. The amount of insulating mud used in the nest's construction might also vary as required.

This bird feeds on a wide variety of insect fare as well as wild fruits and berries, though earthworms are indeed the preferred diet of the robin in summer. These it locates primarily by sight in summer lawns—not by listening for them or "feeling their vibrations" in the soil as formerly suspected.

Until the enactment of the federal Migratory Bird Treaty in 1913 the robin was classed as a game bird in several southern states, and widely hunted for food and sport. Though it is now fully protected as a songbird throughout its range and thus out of reach of the hunter's gun, the annual natural mortality of the robin is estimated at about 80 percent of the population of that year's fledglings, as it is in many other wild bird species.

Blue and Stellar's jays
Cyanocitta cristata and *stelleri*

The blue jay is also known as blue coat, nest-robber, and common jay; the Stellar's jay is known as mountain jay or pine jay.

The blue jay is one of the most active and vociferous—and therefore most apparent—of North American bird survivors. Known to every resident of cities and towns throughout the bird's wide range, the jay lends color, sound, and action to the modern landscape and is generally appreciated for its beauty.

This colorful jay gives a variety of familiar calls, the most common the strident *jay-jay!* that gives its common name. It also utters a harsh, hawklike scream and a querulous, musical *wheedle-wheedle* when communicating with flockmates.

The blue jay occurs throughout the United States east of the Rockies and north to central Canada and Nova Scotia. It is resident throughout range, though northern birds might move south to Florida and the Gulf states during the winter if the weather is unusually severe.

The natural habitat of the blue jay is oak woodlands, but it has adapted readily to the urban and suburban environments, especially where there are plenty of oak and other nut-bearing trees. It is a common resident of city parks and suburban backyards, and a frequent feeder visitor throughout the winter season.

Stellar's jay

STELLER'S

Blue jay

BLUE

QUINN '93

The blue jay undergoes periodic cycles of abundance in the east, with many in evidence one year, few the next. It has been accused of harassing and usurping the territories of smaller birds and driving them away from bird feeders, but these sins are more than paid for by the extroverted bird's lively and beautiful presence in an increasingly monotonous world!

The blue jay's important function as a seed dispersal agent is well illustrated by studies carried out in Virginia, which revealed that in 28 days roughly 50 blue jays transported and cached 150,000 acorns, 58 percent of the total nut crop of 11 large pin oaks in the area. The researchers found that the jays almost invariably selected sound nuts that were likely to germinate when the birds cached them in the ground. Those nuts that fell from the trees and were left on the ground were usually eaten by the jays and other birds or destroyed by insects. The jays apparently tested the plucked acorns for soundness by shaking or rattling them before departing the tree; those discovered to be dried-out or unsound were simply dropped and the bird moved on to another one. Studies of beechnut dispersal by jays has determined that 88 percent of those nuts selected by the birds germinated when retrieved and planted by the researchers, while only 10 percent of those collected by the scientists did so.

The Stellar's jay occurs in coniferous forests in the northern part of its range and in pine and oak woodlands to the south. It is found nearly throughout the West, from coastal Alaska south to southern California and Central America, and east to about the Rocky Mountains. This jay is the only western jay that has a crest. It is a darkish bird, the front half sooty black, the rear dusky bluish-gray.

The Stellar's jay is certainly not the city bird the blue jay is, but it freely invades suburbs and has little fear of humans wherever it encounters them. It quickly exploits the generosity of campers and other human providers and regularly visits bird feeders, winter and summer.

Ring-necked pheasant
Phasianus colchicus

Ring-necked pheasant

Also called ring-neck, pheasant, and partridge, this beautiful Asian fowl is one of the few stories in the annals of animal introductions that has a happy ending . . . sort of. Attempts to introduce the ring-neck into the New World predate George Washington's presidency; reliable records mention several unsuccessful releases as early as 1730. Six dozen pairs were released on Governor's Island by New York governor John Montgomerie, persisting there until the island fell to urbanization in the mid-

1800s. Although repeated attempts were made to introduce the pheasant as a breeding resident of New England, New Jersey, and Pennsylvania throughout the 19th century, nearly all resulted in failure, and for reasons no one could really understand. In nature the ring-neck is far from a tender tropical bird, living as it does in harsh unforgiving environments in its native China. It was thus something of a mystery why the birds could not adapt to the fertile fields and woodlands of New England.

In the end, the pheasant made its first successful landfall on the continent not in the East, but in the Far West. In 1882, Judge Owen Denny, the U.S. consul in Shanghai, imported 30 of the birds from China, shipping them home on the freighter *Isle of Butte*. In March of that year, the ship landed at Portland Oregon, the birds were off-loaded and brought to Oregon's beautiful Willamette Valley, where they were set free on a wooded hillside behind Denny's large estate.

For some reason, the ring-neck found the environment and weather conditions of the Pacific Northwest much more to their liking and the area surrounding the release site was soon knee-deep in pheasants. The birds so successfully and spectacularly colonized their new home that in 1892—a mere 10 years later—Oregon opened its first season on the pheasant and 50,000 were taken on opening day.

The State of Oregon subsequently distributed brood stock far and wide, and as nearly every state was interested in the ring-neck as a potential game bird, the bird quickly became a staple in every area that could support them. The eastern states subsequently imported fresh stock from Europe and Asia, so that by the early years of the 20th century, the ring-necked pheasant was a familiar part of the nation's fauna, except for large areas of the southeast, where spring temperatures can be too high for successful egg development. Today, the pheasant occurs in farming and light suburban habitats throughout most of the rest of the country, and numbers in the tens of millions.

The ring-neck is one of relatively few exotic animal introductions that seems to have achieved success without taking a terrible toll on indigenous wildlife. Although it has increased in number beyond all hopes, the species did undergo a considerable and widespread decline from which it is only now recovering. In the mid-1960s, wildlife managers began to notice that pheasant populations, and hunter bags, were in sharp decline. In South Dakota, one of the top pheasant states, the 1940s population was estimated at between 16 and 40 million birds; the ring-neck was so important to the state's economy that it was named the state bird in 1943. By the early 1970s South Dakota's pheasant population had fallen to about 2 million birds and by 1978 other states were recording similar declines. In Ohio and Indiana populations had fallen an estimated 96 percent and by 85 percent in North Dakota; Illinois' pheasant flock fell to its lowest-ever level in 1984. Throughout the East the bird experienced similar, though not as drastic declines.

At first, the usual culprits were fingered: excessive hunting mortality, unusually severe winter weather, and natural predators taking too large a toll. These potential causes were all closely examined and addressed (some state-instituted ill-advised bounties on predators that did nothing to stem the loss), but it required a long-term look at the problem to discern its true cause.

Gradually, it became apparent that the very industry that had been responsible for the pheasant's roaring success as an avian immigrant—farming—was now responsible for its catastrophic decline. It seemed that as the "conservation ethic" of good farming—that of leaving some fields fallow and hedgerows untouched for wildlife—prevalent up until the mid-1960s gave way to the lure of profits from foreign grain sales, farming as an industry became much more intensive and "cleaner." Farmers cut hayfields earlier, mowed ditches and fencerows, and removed grain stubble right after harvest. All of these practices, along with intensified weed control efforts and more mechanized farm machinery, served to destroy pheasant habitat at a great pace. Where once a farmer's fallow field was a critical winter refuge for the birds, it became a biological desert, unable to support a field mouse, much less a pheasant.

The greatly increased efficiency of large-scale farming served over the past two decades to place the pheasant's future in some doubt, but as of today the trend is showing signs of reversal. The 1985 federal Conservation Reserve Program, which essentially compensates farmers for taking highly erodible land out of production, has seen the recovery of more than 34 million acres of pheasant habitat. In addition, a group called Pheasants Forever, founded in 1982, has purchased some 15,000 acres of prime pheasant habitat in the Midwest and planted more than a half-million acres of food and cover plants for the birds.

Changes for the better in farming practices have also helped the pheasant. Many farmers now sow seed grain directly into the existing stubble rather than plow the entire field and many states now pay farmers to plant "shelter belts" along field edges for the birds. In addition, most states now require hunters to purchase special stamps, much in the manner of trout and duck stamps, before they can hunt the birds.

In the quest to ensure that the Eurasian pheasant remains a part of the North American fauna in the future, another, closely related species was imported and stocked in the Midwest in the late 1980s. The Sichuan pheasant, a darker variety of the ring-neck that also lacks the characteristic neck ring of the latter, has proven to be an even hardier and more adaptable creature. The native habitats of this bird include everything from rice paddies to unforgiving landscapes above timberline at 15,000 feet, and Michigan wildlife specialists hope that the Sichuan pheasant will be able to colonize some 16 million acres of brushland in southern Michigan where the ring-necks have so far failed to gain a foothold. Michigan received 3500 Sichuan eggs from China over a three-year period in the late 1980s and have since liberated 55,000 birds raised in their captive breeding program. Other states, including Pennsylvania, New York, North Dakota, Ohio, and Oregon currently have Sichuan pheasant projects underway.

In sum, the ring-necked pheasant and its several races are tough, resilient and prolific—all qualities that will be found in abundance in any plant or animal survivor.

Turkey and black vultures
Cathartes aura and urubu

Also known as turkey or black buzzard and carrion crow, these big and conspicuous vultures are familiar soaring scavengers over much of North America—the turkey vulture from coast to coast, the black throughout the Southeast. The turkey vulture, incorrectly called "buzzard" (the European term for the soaring Buteo hawks) throughout the South, is found from British Columbia and New Hampshire south to southern South America and the Falkland Islands. It has been slowly but steadily expanding its range northward and has become a familiar sight in the skies of New England, where it was virtually unknown 30 years ago.

The smaller black vulture occurs from about Kansas east to Pennsylvania and south to southern South America; it is the common street and market scavenger throughout the American tropics and the Caribbean, strolling fearlessly among people and their conveyances in the search for edibles. In this respect the black vulture is much more a "people-oriented" bird than the turkey vulture, which nests as far from people and their homesteads as it can manage and generally avoids close contact. This vulture increasingly appears in the skies over the Middle Atlantic States and might soon colonize the Northeast as has the turkey vulture.

Both of these species can be told from soaring hawks by their very small heads and long, broad wings. The turkey vulture holds its big wings in a shallow upward V and flaps much less during flight than the black vulture. It rocks and rolls from side to side while soaring, while the black flaps and soars about more like a hawk or eagle. The turkey vulture appears almost uniformly dark from below, while the black

Turkey vulture

Black vulture

vulture displays two prominent white patches on the outer wings. Both vultures are on the quiet side, issuing only reptilian hisses and low grunts and groans when feeding or fighting. For the most part, the rapid growth of the road and freeway network in North America has benefited both vultures. The great toll of wildlife unfortunately assures them of a reliable food supply, winter and summer. The turkey vulture, in particular, is one of the commonest soaring birds above the vast federal interstate highway system from the Atlantic to the Pacific. Black vultures compete with crows and gulls at landfills throughout their range and are abundant wherever slaughterhouse offal is discarded.

Although the future looks rather bright for these hardy, adaptable and abundant avian trash collectors, the picture could change if extensive areas of rural country go under the bulldozer blade of progress. Both birds are ground nesters that lay only two eggs each season. They require rather large and isolated breeding territories that afford them freedom from disturbance—open and fallow fields and lightly wooded or shrubby terrain. Unfortunately, these habitats often find great favor with housing and industrial park developers, and as these places shrink and disappear, especially at the edges of the spreading cities, so too will the vulture populations, roadkill food supply notwithstanding.

Mallard duck
Anas platyrhinchos

Mallard

Also called stock duck, greenhead (male), wild duck, and gray mallard (female), this wild fowl is without a doubt the most familiar duck, well known to anyone who has visited a public park or a zoo in North America. The original range of the mallard includes most of western North America, from southern Canada east to the Great Lakes and south to the Gulf of Mexico. The species has extended its range eastward over the past 50 years so that today it is a common bird on the Atlantic coast, where it has been gradually usurping the place of its closest relative, the black duck of the vast coastal marshes. Indeed, one of the greatest threats the black duck faces is that of competition and hybridization with the more aggressive mallard.

It was previously suspected that most of the mallards seen east of the Alleghenies were merely escaped domestic stock and their semi-wild progeny, but because many of the individuals seen today are smaller and slimmer wild birds, it would seem that the species has truly undergone a genuine range expansion. A fair percentage of eastern "wild" mallards show enough variations in color and pattern (missing neck rings, patches of white, etc.) to betray their barnyard origins, but increasingly, the general population of eastern birds appear to be of the true wild form.

The mallard is so familiar that it scarcely needs any description; it is the standard against which all other waterfowl are measured and compared for identification purposes. A most attractive bird, this duck often goes unappreciated, lending some truth to the old saying supposedly applied to it: "it would be beautiful if it weren't so common." The male with his iridescent green head and strongly patterned body can be confused with few other ducks, but the brownish, mottled female resembles, to one degree or another, the females of several other species. Consult any one of the comprehensive modern field guides for waterfowl identification.

This duck is a common resident of park ponds and larger fresh and brackish water bodies from coast to coast. It sometimes frequents saltwater, and I have seen females leading broods of tiny ducklings braving the turbulent waters just beyond the surf of the ocean front.

Canada goose
Branta canadensis

Also known as the wild goose, honker, bay goose, and Canada brant, this big bird conjures up images of unfettered wilderness and the lost freedoms of pre-European North America, but in reality, it is today almost as much an urban creature as man himself. Although the truly wild honker is in gradual decline due to hunting and habitat loss, and the subject of some concern on the part of wildlife experts and conservationists, the semidomesticated, or feral edition of the species is undergoing something of a population boom. Particularly in the urbanized East, the Canada goose—like the white-tail deer—has all but ceased being an object of admiration and awe, and become instead an attractive pest.

The Canada goose occurs in six distinct races that vary primarily in size and darkness of coloration. Those found in the Far North and Alaska are very dark brownish, almost black, while the more southern subspecies are paler. Size varies from the mallard-sized cackling goose to the giant Canada goose, which can weigh up to 22 pounds as an adult. The overall natural range of the Canada goose is from the Arctic Ocean south to the Dakotas, and east to the Gulf of Saint Lawrence, with a general winter withdrawal south to the Gulf of Mexico. In many parts of the United States and in particular in the Northeast, the goose has established large and growing populations of essentially nonmigratory birds, the descendants of escaped or liberated semidomesticated stock or true wild birds simply attracted to the reliable food supply and absence of predators. It is these urbanized birds, and not necessarily their wild relatives, that will survive without trouble into the next century.

The feral Canada goose differs in appearance little, if at all, from its wild relative. Unlike the mallard, in which the wild stock is slightly smaller and slimmer and might show color differences, the park honker is about the same size as the wild goose and identical to it in color and pattern. The only real differences are in behavior, for the feral goose associates people with a food supply and seldom strays far from an area unless forced out by intense cold or disturbance.

The goose has been able to gain and maintain its foothold in suburbia for three main reasons: the expansive lawns of public and corporate parks offer superb grazing; there are few if any serious predators; and legions of goose lovers provide an abundant and reliable food source in spite of official pleas to refrain from the practice. In New Jersey, the waterfowl hunting season runs 90 days, but the great majority of the state's feral geese occupy areas that are off-limits to hunters and thus out of reach of the guns of autumn. It is very tempting to speculate that this is deliberate self-protection on the part of the savvy birds.

Although tame geese exist in large flocks in most of the Middle Atlantic states and much of southern New England, they are particularly abundant in the area of metropolitan New York and New Jersey. The latter state's department of fish and game estimates that more than 20,000 birds are now entirely resident there, with the population swelling to some 129,000 during spring and fall migration. Although a great number of the migratory birds seek safe haven in several of the state's wildlife refuges and reserves, most of the veteran people-friendly residents congregate in any place that has abundant grass for grazing, a handy body of water, and cover for nesting—not to mention a loyal following of human goose feeders. Such locations include larger public parks, corporate centers and universities with extensive lawns, abandoned dumps, golf courses, and other such open, seminatural areas.

In such areas the birds might gather in flocks of several hundred, and although many people thrill at the sight of the birds winging majestically overhead in characteristic formation or strolling regally over the grassy sward, others think of the birds' voluminous and sticky excreta whenever they hear the call of the "wild" goose. Indeed, it is the perceived health hazard that has prompted many of the municipalities hosting geese to introduce ordinances against feeding them, though many goose admirers do so anyway. The borough of Morristown, New Jersey annually spends between $35,000 and $40,000 in anti-goose measures, which include "no feeding" campaigns and the installation of reflective tapes, scarecrows, noisemakers, and other ineffective approaches. The state has suggested the placing of colored balloons above areas where the birds congregate and the shaking of the nesting birds' eggs to kill the developing embryo and render them unhatchable.

The general attitude among those urbanites seeking to evict geese from their environs could be summed up in the words of one harried North Jersey man, who complained that geese "are dirty birds. They're nuisance birds and they're only pretty birds when they're flying in a V-formation." If the Canada goose were indeed only passing through, it would doubtless have far fewer detractors in suburbia than it does today.

New Zealand is one of the many places worldwide that the Canada goose now calls home. The big bird was introduced there in stages starting in 1879 and is now established to the point of pest status, along with fellow waterfowl migrants, the mallard and black swan of Australia. The honkers that abound on the island today, devouring grain and competing with sheep for forage, are the descendants of 48 birds released in 1905. By midcentury, the species had been taken off the protected list and classed as a game bird in order to control their growing numbers.

Mute swan
Cygnus olor

This graceful, beautiful, and often aggressive bird stands alone among North America's three swan species as the one likely to persist well into the next century, at least in the semiwild and free state. The mute swan, native to Europe and the British Isles, was introduced as an ornamental bird on several estates on Long Island, New York, in the 1880s. Unlike the ubiquitous house sparrow and starling, the swan's rate of increase following escape or liberation was very slow. It has gradually extended its range so that today it is seen up and down the Middle Atlantic seaboard from Massachusetts to the Chesapeake Bay, with a smaller, disjunct population established in Michigan. The swan is most common, however, in the environs of New York and New Jersey, where large numbers breed and winter in larger parks and reservoirs and on coastal ponds and bays.

The mute swan is the only swan found in its range, and thus seldom competes with the native whistling or tundra swan, or the endangered trumpeter swan—both arctic and western species that never come into contact with the European interloper. If there were serious competition for food and nesting sites among these species, the two native swans would doubtless lose out, for the mute swan is a very aggressive, fearless bird that will not hesitate to attack man or beast in defence of mate and young. There have been many reported incidences of swans attacking and severely injuring small dogs and very young children when the birds are approached while escorting cygnets. This swan is so territorial and belligerent while nesting that other waterfowl are very often attacked and driven out of the claimed area, which can include an entire park lake.

The mute swan is just that—almost entirely silent except for hisses and low grunts when angry or threatened. Their great wings, however, give vent to a loud swishing sound when the birds take off from water or are flying overhead. The combined splattering noise of the large feet and whistling of the wings makes watching a flock of swans become airborne a memorable experience.

Chapter seven

The mammals

The class Mammalia is the smallest of the world's vertebrate divisions, containing a mere 7400 known species, as opposed to roughly 32,000 fishes, 8500 birds, and 11,000 reptiles and amphibians. Among the insects and other invertebrates, by far the largest group, there are more than 20,000 known species of beetles alone.

In one way the mammals are a paradox. Though for the most part large, active, often noisy, and conspicuous as living entities, the great majority of those species native to North America are secretive and nocturnal. They carry out their affairs well out of sight of humans, who are essentially diurnal (day-active), and tend to shun the crepuscular pathways that most mammals prefer to travel.

Many of the world's mammal species are in jeopardy in the waning years of the 20th century. The list of threatened or endangered mammals mostly consists of the larger and medium-sized carnivores and herbivores—animals like elephants, okapis, rhinos, wolves, and grizzly bears, which simply cannot exist without an abundance of territory. They are also vulnerable to being shot on sight, either as food for man or because they are in competition with him for space, or the large tasty herbivores that occupy it.

On the continents of Africa and South America, a diversity of large and small mammals still exists in dazzling variety and beauty. However, as the Third World nations that occupy those land masses aggressively pursue economic, agricultural, and industrial development—not to mention spectacular population growth—in the next decade, most of the larger mammals will be driven to extinction in the wild state and will exist only in the biological limbo of rigidly protected preserves or zoological collections. The American bison, the white rhinoceros, and the Przywalski's horse are apt examples of species sharing that fate. The outlook for the white rhino, for example, is particularly grim. Continued illegal poaching in Zimbabwe has reduced the already desperately low 1990 population of 2000 animals to 500 in late 1992. In Kenya the black rhino, estimated at 18,000 animals in 1968, has shrunk to a mere 400. Surely the rhinoceros is doomed as a free-living creature and viable species, and well before the turn of the new millennium.

The tiger is another case in point. An estimated 40,000 tigers lived in India at the turn of the century, whereas unregulated hunting and habitat destruction have reduced that number to less than 2000 today. Tiger specialists with the International Union for the Conservation of Nature and Natural Resources say "the end of the tiger is in sight, possibly within ten years."

Other, smaller mammals are disappearing for a host of other reasons, not the least of which is the nearly universal threat—habitat destruction. The mammalian fauna of any ecosystem can be decimated by two means, "indirect defaunation," or "direct defaunation." The latter involves the direct killing of animals for food, sport, byproducts such as leather or fur, or predator control and does not result in the alteration or destruction of the habitat except where burning might be practiced as a means of flushing victims out.

Direct defaunation can be illustrated by a study conducted in the Brazilian Amazon that showed that a single family of rubber tappers in a year and a half killed for food more than 200 woolly monkeys, 100 spider monkeys, and 80 howlers. The commercial wildlife trade also takes a horrific toll, as exemplified in statistics for animals shipped from the Amazonian port of Iquitos between the years 1962 and 1967. The mammalian victims included 183,664 live monkeys; the skins of 67,575 capaybara; 47,851 otters; 61,499 ocelots; 9565 margay; 5345 jaguar; 690,210 collared peccary; 239,472 white-lipped peccary and 169,775 "deer." These figures, added to those of other assorted wildlife trans-shipped from the port over those three years led researchers to compute a total of five million animals killed, or approximately 800,000 animals per year. This would indicate the loss of at least one larger animal for each square kilometer of the entire Peruvian Amazon drainage each year.

Indirect defaunation is more insidious but nonetheless equally devastating, and results from the rate of deforestation currently extant in the tropics today. About three billion acres, 11 percent of the earth's vegetated surface area, have become eroded or desertified from deforestation and overgrazing since 1945. All of the activity that has resulted in this environmental horror-in-progress was not directed at animals specifically, but rather at their habitat. As the natural forest cover is disturbed or removed through activities such as subsistence farming, firewood gathering, mining activity, or logging, "islands" of undisturbed, intact moist woodland remain, seemingly healthy and filled with life in the form of soaring tropical trees and an abundance of bright flowers.

But looks can be deceiving, as investigators have found. Selective logging, subsistence hunting, the collecting of fruit and nuts for sale, and a host of other human activities both inside and outside the intact forest has removed entire populations of larger mammals and birds from the otherwise healthy habitat over vast tracts of rain forest in equatorial countries. Subsistence hunting and collection for the animal products (skins) trade in Amazonian Brazil has resulted in the estimated loss of some 19 million animals yearly, with that figure rising to a possible 57 million if escaped but fatally wounded animals are taken into consideration. In most cases, exploiters of the forest fauna prefer the largest animals for either their food value or pelts, leaving great areas of the habitat void of important links in the food chain and natural plant generation and succession.

Empty forests are becoming an increasing concern in the tropics, but what about North America? Here, roughly 96 percent of the original old-growth forest has long since been removed, the remaining area of truly original forest cover situated primarily in the Pacific Northwest and western Canada. The fauna of second- and third-growth forests—including those that are managed for sustained timber harvest today—has shown a gradual but noticeable decline over the past 50 years.

As the higher vertebrate species dwindle and vanish in the wild state, humankind, whether out of a sense of guilt, regret, or an innate reluctance to stand by and let it happen, has taken steps to rescue a small number of the endangered ones through captive breeding programs. Most of these efforts are being carried out in zoological parks and private collections worldwide, and the main focus of the programs are those highest of vertebrates—the mammals and birds.

Currently, about 150,000 specimens of nonfish vertebrates are housed in North American zoos. Of mammals, there are at present approximately 50,000 mammalian specimens of some 800 to 1000 species exhibited in zoos. This allows for an average of 50 individuals of each species in the 150 North American public zoos, though relatively few of these animals are being managed as part of organized captive breeding programs. The general zoo population is primarily composed of the ungulates, or hoofed mammals (40 percent), primates (24 percent), and the carnivores (18 percent).

These creatures are highly visible, and thus more easily observed and enjoyed by paying zoo visitors. Oddball creatures, such as the secretive bats, comprise the remaining 18 percent of the zoo roster, though they represent fully 20 percent of living mammal species.

The mammal survivors of tomorrow, of course, include those large, genetically-engineered herbivores—domestic cattle, sheep, goats, horses and hogs—that have served as humankind's laborers and meaty fodder for millennia. These creatures will not only be with us in the captive state for the foreseeable future, but are one of the principal reasons why so many species of wild creatures are threatened today and won't be around for much longer.

Livestock numbers worldwide have increased dramatically; in fact they've doubled right along with the human population since 1960. There are now roughly 1.264 billion cattle in the world, about two-thirds of them in Third World countries. About 1.173 billion sheep are busy turning grasslands into desert, about half of them in developing nations, the other half in the industrialized countries. Hogs are estimated to number 823 million; goats about 520 million, nine-tenths of them in the Third World. There are about 122 million horses, mules, and asses, 98 million of these in the developing nations. The pasturage required by the world's growing livestock throng is one of the major causes of deforestation and the continued destruction of natural grasslands today.

Given the rapidly changing character of the world's natural environments, what are the qualities conducive to survivorship in those mammals that will inhabit them in the future world? What constitutes an adaptable, "r-selected" mammal species, as opposed to one doomed to extinction as people appropriate its habitat and thus directly or indirectly affect its existence?

The answer to this question can be found, for the most part, through an investigation of a given species' two main physiological criteria: body size, and behavior. It is pretty clear that a large, diurnal, wild mammal—herbivore or carnivore—is going to have a difficult time avoiding people if they are present in its habitat in any numbers. Elephants, bison, wild horses, lions, and the like all move about primarily by day. They are conspicuous and often aggressive animals in their respective environments. In addition, they require large amounts of territory in order to function as viable organisms, and their feeding habits and disregard for human property lines often bring them into direct conflict with people.

Medium-sized nocturnal mammals, such as the red and gray foxes, the coyote, the mountain lion, and the black bear might hang on with varying degrees of success under persecution and severe habitat loss, but once the human population density of an area reaches a certain level, most will surely decline and disappear. Only the raccoon, the opossum, the skunks, and the coyote, to name a few, have managed to persevere in heavily suburbanized environments with any real degree of success. Larger mammals, even night-wanderers, need correspondingly larger territories, and as more and busier roads bisect woodlands and prairies everywhere, many wild creatures are killed attempting to cross them. The cover of darkness will cover an animal's activities quite well, but a big raccoon rummaging through the trash makes a lot more noise than does a house mouse, and sooner or later the animal control officer is called in to remove the "nuisance animal."

Smaller, nocturnal mammals, on the other hand, can easily avoid direct contact with people by virtue of their size and habits, which permit them to occupy "personal spaces" below the notice of the much larger humans, and to travel at a time when the essentially diurnal Homo sapiens is traditionally catching forty winks. For the most part, this group consists of the huge rodent tribe, as well as the rabbits and hares and such reclusive creatures as weasels, shrews, and moles. It is among these mammal groups that the majority of tomorrow's mammal survivors will be found.

Feral dogs
Canis familiaris

The dog is the domestic animal that has had the longest and most intimate association with mankind. The relationship extends back into the Stone Age, long before people brought such wild creatures as sheep, pigs, and cattle into his agricultural fold. In all likelihood, the dog's origins lie in the prehistoric wolves that took to following nomadic human hunters in hopes of securing a share of the spoils of his kills. As time went on, these man-associated wolf populations underwent both physical and behavioral changes that made them more receptive to a closer, more intimate relationship with man the hunter—that of home companion and direct partner in the hunt.

It is known that at least four distinct dog breeds existed in Eurasia by 8000 B.C., and Native Americans had developed a number of working breeds by the time of Columbus's arrival in the New World. The man-dog relationship then was a strictly utilitarian one. Canids were used primarily as beasts of burden, hunting allies, and as a source of meat and furs; this purely working relationship survives today in arctic Eskimos and other Asiatic peoples, who seldom, if ever, make pets of their dogs in the Western sense of the word.

There could be as many as one dog for every three people in the United States today. The Pet Food Institute estimated in 1990 that there are 50.5 million dogs in 33.5 million households in the United States. This figure does not include the hundreds of thousands of strays and half-wild feral dogs roaming the continent, and is very likely conservative; zoologist and canine specialist John C. McLoughlin places the total number of dogs in the United States at 80 million. It is estimated that somewhere around 2000 puppies are born each hour in the United States. Of this great throng of animals, some 10 million are yearly put to death in municipal and humane society shelters; the City of New York alone destroys between 30,000 and 40,000 animals a year, the great majority of them dogs and cats. As an indication of the food required to nourish this growing army of pet dogs, it is estimated that in 1990 American dog owners bought 6.6 billion pounds of dog food at a cost of some $3.3 billion.

No other domestic animal displays the degree of morphological variation found in the dog. A cat, no matter the breed, is still recognizably a cat, and the primary differences among breeds will be found mostly in color, pattern, and hair length. All adult cats of every breed, be it the naked Sphynx or the lushly-haired Persian, are essentially the same size, and their facial confirmation clearly says "cat."

In the dog, however, the physical differences are so great as to cause wonder about whether some of the most disparate breeds in fact had a common ancestor. Today's *Canis familiaris* ranges in size from such large breeds as the mastiff, which can weigh in excess of 200 pounds, to the miniature Chihuahua and Yorkshire terrier, both logged in at less than 16 ounces when adult. The very smallest adult dog ever verified was, according to the Guiness Book of World Records, a miniature Yorkshire terrier that stood an amazing 2½ inches at the shoulder and was 3 ¾ inches long from the nose to the base of the tail. This lilliputian creature, a dog nonetheless, weighed an unbelievable four ounces!

When you compare a stubby-legged dachshund with a gazellelike greyhound, the potential variation in physical conformation in the dog is immediately apparent. These two breeds, though both descended from the original wolf stock, cannot breed naturally; they would have to be cross-bred through artificial means. Likewise, the skull of the achondroplastic Pekingese, which suffers from a condition that affects the growth of bone cartilage, bears no resemblance at all to that of the wolf. Whereas the wolf skull is sturdy, long-profiled and armed with the prominent and formidable teeth of the predator, that of the Peke is fragile and bluntly rounded, with ineffective teeth only partially rooted in the weak jaws. This stunted toy dog must thus be fed presoftened foods if it is not to starve to death.

In between all of the some 200 refined working, sporting, and toy breeds are the "mixed breeds," or mutts. Allowed free rein in the romance department, all dogs will, of course, freely cross man-enforced genetic boundaries, and the resulting crosses always show more physical vitality and fewer congenital defects than the purebred strains. One of the most typical of the "mutt-type" dogs is the so-called "pariah dog" of Asia and Africa. The word pariah stems from an Indian word meaning "drummer," which was usually applied to those people of the lowest warrior castes, those fit only for the beating of drums in battle. The pariah dogs, often scorned and even feared as livestock predators wherever they roam freely, are reportedly the result of deliberate cross-breeding at some time in the distant past. Their ancestors were likely both wolves and wild dingo dogs.

The modern domestic dog, pure bred or mutt, can create big trouble for human society if it steps out of line, as anyone knows who has ever been threatened by a pack of free-roving dogs, or who has witnessed or read of attacks by dogs on family pets, deer, or domestic livestock.

New York City's dog population deposits an estimated 5000 to 20,000 tons of feces annually, and in 1970 there were 6809 reported dog bite cases in Baltimore alone. The problem is not one of recent origin. In 1722 the City of Philadelphia authorized its residents to "kill on sight the authors of any canine disturbances."

The domestic dog easily slips into the feral (meaning "domestic gone wild") state, quickly hooking up with other strays and adopting the social, pack-hunting behavior of the ancestral wolf. Although feral dogs can and do breed in the semi-wild state, only in the more southern parts of the continent, where winters are less harsh, do they have any real chance of maintaining stable breeding populations. Unlike the wolf and other wild canids, which undergo estrus and whelping at specific and favorable times of the year, the domestic dog can come into estrus and breed two or even three times a year, regardless of the season or weather conditions. Thus, puppies born in the middle of a New York State or Wisconsin winter will have little chance of survival, and most northern feral dog packs shrink and vanish after a time, their ranks thinned by disease, malnutrition, and periodic assaults by animal control agents.

Feral, pack-living dogs are primarily a phenomenon of the inner cities and extensive rural areas rather than the burgeoning suburbs. Such large carnivores, moving about during the day in groups—and generally creating trouble wherever they go—quickly attract attention in suburbia, and are summarily rounded up and disposed of. In the relatively wide-open spaces of the rural lands at the edges of the cities, roaming dog packs have much more room in which to wander and elude capture. The dog's inherent intelligence and degree of social organization greatly aids in the survival of feral dog packs, many of which become active hunters of deer and pastured domestic animals.

The urbanized world in which the modern feral dog exists is, in the words of zoologist McLoughlin, "a twilight econiche" in which competition with cats and rats for food renders their lives "nasty, brutish and short."

In the cities feral dogs survive primarily on refuse and on the carcasses of other animals found alive or dead. Overall public indifference to their presence combined with the formidable logistics of hunting them down in the complex maze of buildings, streets, and decaying industrial areas that are the modern urban landscape has made the elimination of urban feral dog packs a costly and frustrating task. New York City alone pays the American Society for the Prevention of Cruelty to Animals $4.5 million a year to pick up stray animals, the great majority of them dogs. In 1992 the Society—doing a job it says it no longer wishes to do in the interest of improving its public image—bagged 56,000 animals, the bulk of them dogs. Of this total, nearly 10,000 were placed in adoptive homes and the rest were euthanized.

Feral cats
Felis domesticus

The domestic cat is one of the most abundant small carnivores in North America today. Just how numerous they are can be seen in the startling statistics that nearly six million unwanted cats are euthanized each year in the United States, while some 35,000 kittens are born each day. This birth rate far outstrips that of humans, of whom a mere 10,000 enter the world within a 24-hour period. Although a fair percentage of these animals are house pets under at least some degree of supervision by their owners, a vast number of house cats are feral, living in the "wilds" of agricultural areas, suburbia, and cities, preying on the resident wildlife.

The pet (not total) cat population of the United States in 1991 was estimated at 57.9 million animals in 27.7 million households, making the cat somewhat more popular as a house pet, at least in numbers kept, than the dog. The cat's numbers, replenished with reckless abandon yearly, currently equal about 30 percent of the human population of the U.S.—exclusive of the armies of feral strays prowling the cities and farms, that is.

As a species, *Felis domesticus* has shown less malleability for genetic tinkering—and thus man-engineered variation—than the dog. The largest domestic cat breed is the so-called "rag doll," which averages some 15 to 20 pounds at maturity. The Singapura, or "drain cat" of Singapore, weighs in at a mere six pounds. The average well-nourished tabby weighs about seven pounds in the female, a bit over eight in the tom.

While all cat breeds are readily recognized as cats no matter how different they are, the various breeds can vary considerably in overall appearance. For example, contrast the snub-nosed, long-haired Persian with the slender, saturnine, utterly hairless Sphynx cat. Unlike the more over-bred varieties of the domestic dog, nearly all cats can be efficient hunters when they want or have to be. Even a pampered Siamese or Persian can catch an unwary sparrow if it is hungry or tries hard enough, for selective breeding has not removed all of the innate grace and supple power of the original wild feline ancestor.

A four-year study conducted by researchers at the University of Wisconsin, Madison, determined that free-roaming cats might annually kill 19 million songbirds and 140,000 game birds in that state alone. The study was conducted by radio-collaring 30 farm cats for various periods of time and following their hunting activities.

A 1987 British study revealed that England's five million cats annually account for about 70 million small animals, 20 million of which are birds. The study, conducted in a small village in Bedfordshire, England, determined that the yearly toll of the 78 cats observed included small animals such as mice, voles, and shrews (64 percent), and small birds such as house sparrows and thrushes (36 percent). The researchers, estimating that the average prowling pet cat will bring home about half of its nightly victims as "trophies," asked the village's cat owners to retrieve the catch and stash it in plastic bags. After a year it was found that the cats had killed at least 1094 small animals and birds.

Contrary to popular opinion, the house cat is not an overly effective hunter and destroyer of rats. Many cats are in fact fearful of tackling the feisty and aggressive rodents. Studies made in Baltimore have shown that feral cats and rats often coexist with relative harmony, the two species virtually ignoring each other as they mutually forage among the urban trash for their nightly sustenance. In more than 1000 hours of observation, the investigator was able to witness only five incidents of cats killing rats. The Baltimore study also showed that both cats and rats esteemed garbage as a source of food, with both animals having a seeming preference for discarded scrambled eggs, macaroni, and various cheese items.

Feral cats are much more numerous in the environs of farms, suburbs, and cities, where virtually all of the natural predators have been removed, than they are in untrammeled wilderness where food is scarce and predators more abundant. The most serious enemies a wandering suburban cat confronts are dogs, hostile humans, and the private car. The latter annually dispatches some 1.5 million cats on the streets and roads of America.

Cat density per square mile can vary widely, according to the habitat and the availability of food in the form of waste foods or small prey animals. In general, there will be many more feral cats within urban areas, where buildings and other constructions offer a maze of hiding places and den sites, plenty of prey animals, dedicated cat feeders, and edible garbage.

Muskrat
Ondatra zibethica

This medium-sized rodent is one of the most widespread and yet least known of our small mammals. The muskrat occurs over a vast area of North America, from Newfoundland and Labrador west to the Northwest Territories and Alaska, and south to Georgia, Texas, and Mexico. It is absent in peninsular Florida, where its niche is occupied by the Florida water rat (Neofiber alleni). The muskrat's principal environmental requirement is the presence of water, and thus this animal is primarily a marshland creature.

The muskrat, or just plain "rat" to fur trappers, reaches a head and body length of 10 to 14 inches, and weighs between two and four pounds. It is distinguished by its dense, rich brown fur and the laterally flattened tail. The rudderlike tail is an aid to navigating the aquatic and swampy habitats this animal favors, and the 'rat can swim and dive with the best of them. In years past the muskrat's fur commanded high prices—as high as $10 per pelt—on the fur market, but in recent years it is in much less demand. This is due in part to the effective animal rights protests of recent years, which have resulted in a lowered demand for even inexpensive furs. Moreover, far fewer boys and young men, even rural youngsters, are inclined to try their hand at earning extra money by "rat trapping." The demands of the fur trade are at least one peril the fecund muskrat doesn't have to contend with much anymore. This aquatic rat is still trapped and eaten extensively in the South, however, particularly in coastal Louisiana.

Although this animal must have water nearby in order to set up housekeeping, it is by no means unreasonably specialized in that area. Where muskrats inhabit freshwater and tidal marshes, they build their characteristic conical lodges of mud and marsh vegetation. In park ponds and along even heavily urbanized rivers and drainage ditches, they simply burrow into the mud banks, driving their complex system of tunnels and resting dens deep into the soft ground. Where muskrat populations are high the animals fan out and occupy any suitable habitats offering them a small area of water and enough green vegetation to eat. They will be found living in suburban drainage culverts, ditches, and canals, and are commonly observed going about their business in wet spots on superhighway median strips and alongside huge shopping malls.

Muskrat

Muskrats are among the commonest mammalian residents of the great, urbanized marshlands of the industrial Northeast, in particular those vast plains of phragmites and toxic mud of the New York Metropolitan area. Muskrats are often seen foraging alongside the New Jersey Turnpike during the night hours. Perhaps as a testament to the animals' urban sagacity, few are foolish enough to try and cross this wide busy roadway in search of greener pastures on the other side, and thus the sight of muskrat road kills is relatively rare on the Turnpike.

Another large aquatic rodent that has shown signs of range expansion and ultimate mammalian survivorship is the South American nutria. Somewhat larger than the muskrat, with a rounded hairy tail as opposed to the naked, vertically flattened tail of the muskrat, the nutria was introduced into Louisiana in the 1870s as a potential fur-producer. Due to escapes and liberations over the years, the nutria has spread throughout the vast marshes of much of the Gulf Coast area and has become established in parts of coastal Oregon. It has been reported as far north as Michigan in the central states.

Meadow vole
Microtus pennsylvanicus

Voles are small, mouselike mammals found wherever there is adequate grass cover. They are found in abundance in all of the continental United States except three states—Mississippi, Alabama, and Florida.

By far the most widely distributed vole is the four-inch meadow vole, found from arctic Alaska and Labrador south to Georgia, Texas, and California. This tiny creature is active by day and night, roaming the convoluted runways and tunnels of its grassy domain, which might be in a wide farmer's field or a suburban lot. This vole varies from gray faintly washed with brown in the western part of its range, to dark brown in the East. The belly is silvery to buff, and the tail is whitish below with a dark gray upper surface. This vole can weigh 70 grams if it is a heavyweight.

The secret of the voles' survival lies in their unobtrusive way of life; few people are aware of their presence during the warmer months of the year, though in winter their round burrow openings and meandering trackways in the snow give them away. The food of this little creature consists primarily of tender vegetation, nuts, seeds, bark, lichens, and insects. During the winter they will strip and eat bark and buds, and they can do some damage to the root systems of ornamental plants and crops.

One to eight young are produced in a well-hidden nest of grass and shredded vegetation after a gestation period of 17 to 20 days.

House mouse
Mus musculus

Although the house mouse is every bit the noxious rodent pest the rat is, the little creature seems to have fared somewhat better than its bigger and more aggressive relative in its public image. While rats have traditionally been assigned the role of sneaky villain or unsavory haunter of dark and dank places, the mouse, though it is pretty much both of these things on a smaller scale, has enough endearing qualities to make its presence at least tolerable, if not desirable, to many people who would otherwise tremble with apprehension at the sight of a large rat in the kitchen. The most famous make-believe animal in the world is a mouse, and no rat, real or imagined, can even approach Mickey's universal appeal.

The word *mouse* itself derives from the Latin *mus*, which stems from the Persian *moosh*, both of which trace their origins to the Sanskrit word, *musha*, all essentially meaning "to steal", or "little thief." The mouse is mentioned as a pest of mankind in the Bible in Leviticus (Chapter 29, Verse 11) and was well known to virtually all of the ancient civilizations as a low-profile camp follower and raider of kitchen provender.

The house mouse originated in Central Asia. It was soon transported by human agency to India and thence to Africa and the Mediterranean region, where it had apparently arrived in the company of wandering farmers by 8000 to 4000 B.C. Today, the mouse is found virtually worldwide, wherever people live and inadvertently provide it with the food and shelter it requires. No one ever deliberately transported house mice from one place to another as desirable animal companions. Most traveled by stowing away on ships or hiding in land-borne shipments of grains and other foodstuffs.

As numerous as the feared and despised brown and black rats are today, there are many more mice than rats on this continent. If there are in fact more than 150 million rats in North America (probably a conservative estimate), then there are many more than a billion mice. In 1926, a single haul of dead house mice poisoned on a single California ranch consisted of eight wagonloads, or more than eight tons of the little rodents.

Although the domesticated albino form of the brown rat makes an acceptable—even affectionate—pet, the "white mouse" of pet shop fame is much more commonly kept by those people interested in deliberately bringing any sort of rodent into their homes. Its smaller size and nervous but nonthreatening behavior doubtless make it preferable as an off-beat pet to the larger and still vaguely dangerous rat, no matter how tame and cuddly individual pet rats might seem. The white mouse is an enormously popular pet, and perhaps the number one laboratory and experimental animal, but studies have shown that its deductive and reasoning powers are exceeded by the wild deer and white-footed mice, which display greater curiosity and motivation than their domesticated and inbred brethren.

Domesticated house mice have been used in successful experiments concerning terrestrial mammals breathing underwater like fishes. The mice were submerged in water supercharged with oxygen and containing salts close to the levels present in the animals' tissues. The State University of New York experiments determined that the animals could survive underwater for long periods of time, though all of the mice died when removed from their temporary liquid home. The work was seen as a possible alternative to deep-divers using a gaseous mixture of nitrogen and oxygen for dangerous deep dives and for avoiding the "bends" and "rapture of the depths."

Sanitation campaigns involving the removal of garbage and trash that offer mice ample hiding and breeding places, combined with intensive poisoning and trapping programs, can make inroads into local mouse populations. But it is very doubtful that the house mouse will ever be fully eliminated from our towns, cities, and farms within the foreseeable future. The species is just too adept at learning from experience and avoiding traps and poisoned baits. Moreover, its astounding rate of procreation assures that even if there are but two survivors of an eradication assault, and they are of the opposite sex, they will at once begin to bring the mouse numbers back up to normal.

White-footed and deer mouse
Peromyscus species

Between them, the ranges of these two mouse species cover most of North America with the exception of the treeless, Far North tundra regions. The white-footed (*P. leucopus*) and the deer mouse

(*P. maniculatus*) both occupy a wide variety of habitats, from prairie to woodlands and brushy areas, and are frequently common in suburbs where they might enter houses in the fall. These mice, which can be amazingly abundant in good years, are tiny creatures seldom exceeding a body length of four inches. Both range from a pale tan through fawn color to a rich reddish brown (depending on the region and the habitat) and white below, and can be difficult or impossible to distinguish as they dart swiftly through the vegetation. In general, though, the deer mouse is slightly smaller than the white-foot, and always shows a bicolored tail, which is longer than the body and darkish above, clear white below. The white-footed mouse has the shorter tail of the two species, usually shorter than the length of the head and body. These two fieldmarks are not, of course, readily noted unless the animal is in hand, a state of affairs not normally encountered by the casual mouse-spotter. Both species are rather big-headed little creatures with the large, bright eyes of the nocturnal wanderer. They differ greatly from the plain grayish, relatively small-eyed house mouse wherever the species occur together.

Deer mouse

White-footed mouse

Both species can be active throughout the winter. Their tiny tracks might be the only mammalian signs you see in a snowy suburban woodlot, other than those of the housecats that might try to bag one of these agile and wary little animals. In spite of a vast array of natural and feral predators, these mice are able to maintain their numbers as long as food supplies are adequate and enough natural cover is present. The little mouse mothers are devoted to their progeny and will go to any length to protect them, even standing up to much larger animals when the brood is threatened. I once discovered a white-footed mouse living with her brood of eight half-grown young in a bureau drawer in a rental cabin on Cape Cod. The tiny creature, although trembling with fear, fixed me with a resolute and bright gaze and refused to vacate the nest; closing the drawer softly, I went to get a camera, but on reopening the drawer a few minutes later I found that the little mother had left the premises, taking all of her youngsters with her.

White-footed and deer mice are generally considered unobtrusive and harmless members of our mammalian fauna, but the Center for Disease Control has determined that these rodents might be responsible for the spread of a new virus that has caused human fatalities in the Southwest. The virus, a new strain of hantavirus, has killed 19 people and caused severe reaction in 34 others. It is believed to be spread by contact with deer mice or through exposure to airborne particles of rodent droppings.

Norway and black rats
Rattus norvegicus and rattus

The Norway or brown rat is a native of Asia. For many millennia it remained there as a natural and integral part of the fauna, with its own complement of predators and controls. The animal's residency on the North American continent dates almost exactly to the year of the founding of the Republic. It is suspected that the creatures arrived here as stowaways in large containers of grain the Hessians shipped here to feed their horses during the Revolutionary War. In 1776, the brown rat was already a resident, albeit very much a minority, of the New York-New Jersey area.

The brown rat is now found in every state and province with the exception of far northern Alaska; it is reliably estimated that this animal has now invaded—via passage by ship—about 85% of the world's islands, with dire consequences for native insular wildlife.

The name "Norway rat" is a misnomer, for this animal hails originally from Eurasia and was certainly not restricted to Scandinavia at any time in its biological history. The black or wharf rat is of more southerly distribution, being most common from southern New England south to Florida, Texas, and the West Coast. It occurs throughout Mexico and Central America.

The brown rat has been responsible for some of the greatest plagues in history. The Black Death (1345–47) killed some 34 million, one out of four people, in Europe. More recent rat-caused plagues have included two massive outbreaks in India: the 1898 plague killed 5 million people, and another in 1907 resulted in 1.2 million deaths.

The brown rat reaches a maximum length of 19 to 20 inches, but the average is 14 to 16 inches, including the scaly tail. It weighs in from 8 to 24 ounces, the heaviest recorded specimen tipping the scales at one pound, nine ounces. The black rat is a slightly smaller and slimmer animal; it occurs in both a brown and black phase, differing from the brown rat in that its tail is longer than the head and body, and is smooth and scaleless.

The rat's teeth grow at a prodigious rate during its two- to three-year lifespan, but through constant wear by chewing the animal keeps them well worn. It has been calculated that if such abrasion of the teeth did not take place naturally, the teeth might attain the length of 30 inches over the creature's lifetime, assuming the beast could survive such a deformity!

The rat is a prolific creature. An adult female might produce almost four litters every three months under favorable conditions, totaling some 15 litters per year. As each litter contains an average eight young (though up to 19 have been counted), the scope of a particular rat problem can be considerable. Young rats are completely weaned and on their own in a bit more than three weeks; shortly after that they can breed.

Cats make only so-so ratters; given an adult rat's size, boldness, and undisputed fighting abilities, some cats are distinctly afraid of them and will not go out of their way to provoke a confrontation.

Gray squirrel
Sciurus carolinensis

Believe it or not, the gray squirrel used to be in big trouble in North America. In the latter years of the 19th century, large-scale, highly destructive logging was going on all over the eastern half of the country, and population centers full of squirrel lovers and feeders were a thing of the distant future. The bright and resourceful gray squirrel was hard pressed to stay ahead of the ominous threats arrayed against it. In those days, the little animal found itself caught in a sort of environmental limbo: it's natural habitat was disappearing at a fast clip all around it, while the artificial one that would serve it so well in the 20th century had not yet come to pass. In the year 1900, the gray squirrel was thought to be an endangered species, although that grim, all-too-familiar term of today had yet to be coined.

The gray squirrel is a very abundant animal today, but its numbers were far greater in the 1700s, before the primeval forests had been stripped from much of eastern North America. Throughout the 18th century and for most of the next, the squirrel was shot in the hundreds of thousands as a garden pest and for the pot. In 1749, Pennsylvania paid out squirrel bounties totalling 8000 pounds sterling, and in 1834 two hunters shot 900 and 783 squirrels, respectively, in one day. It would seem that long before human constructions like shopping centers and superhighways became a threat to the squirrel's welfare it was on its way to disappearing.

The gray squirrel is an arboreal creature needing trees for their shelter and abundant food crop for survival. As befits the true survivor, the hardy trees lining the streets of suburbia, or even those of our largest cities, are a perfectly acceptable substitute for the untrammeled wilderness it once called home. In fact the little "tree rat" prefers it that way. It is by far the most familiar small mammal east of the Mississippi, the great majority of people having nothing but affection for it wherever it occurs in close association with man in his parks, suburban backyards, and even in the environs of our largest cities. It has been introduced and is spreading in several localities on the West Coast, especially in Washington State.

The gray squirrel's scientific name, *Sciurus*, offers a hint as to its overall appearance, for it stems from the Greek words *skia* and *oura*, meaning "creature that sits in the shadow of its tail," or, simply, "shade-tail." The name *squirrel* is from the French *esquirel*. The animal is a true-blue native North American in every respect, occurring in three distinct subspecies from Maine south to Florida and west to Saskatchewan and Texas. The eastern gray squirrel (*Sciurus carolinensis*) occurs from northern Maine south to Florida, and west to Minnesota and central Texas. The western subspecies (*S. griseus*), is found from Washington State south to northern Baja California. The Arizona gray squirrel (*S. arizonensis*) occurs in the oak and pine forests of southeastern Arizona.

The great American deforestations of the last century led to the squirrel's exportation out of danger by well-meaning souls in order to rescue it from possible extinction here, and it has now become a pest of moderate dimensions in England. The first known importation made in the gray squirrel invasion of Britain were 10 animals carried "over there" in 1890 by one G.S. Page of New Jersey. The animals were presented to the Duke of Bedford, who happily set them free on his huge estate at Woburn. Over the next two decades some of the progeny of these original immigrants were caught and distributed throughout southern England; today the gray squirrel occupies virtually all of England, Scotland, and Wales.

Both the gray and the smaller red squirrel (*Tamiasciurus hudsonicus*) were well known to native Americans, supplying them with both meat and fur. The Cherokee name for the gray squirrel was *saloli*. The Indians were loathe to eat the animal because its characteristic hunched position while eating was thought to aggravate rheumatism.

The average adult gray squirrel weighs just over a pound, though winter adults might look heavier due to the lush coat. It can reach 20 inches in length, including tail. Gray squirrels occur in several distinct color phases, from melanistic (black) to albino, and many shades of russet and gray in between. A large concentration of albino grays found in Olney, Illinois, has been traced back to a single pair owned by a saloonkeeper in 1892.

Most tree squirrels, the gray in particular, are known for their periodic mass migrations, thought to occur to relieve the stress of overcrowding following boom squirrel years. The earliest accounts of such movements, often involving literally millions of squirrels, date from 1749 and 1780, indicating a cyclic occurrence. A number of great migrations were reported in New York State and New England throughout the 1800s, but one of the largest of them all consisted of an estimated half-billion animals on the move in southern Wisconsin in the fall of 1842. This enormous assemblage was reported to have been 150 miles long and 130 wide and looked, according to an eyewitness, "like an army on the move."

One of the more recent mass migrations occurred in 1968, when some 20 million gray squirrels took to the road in most of the eastern states from Vermont to Georgia. The driven creatures were killed in the thousands as they crossed lakes, rivers, and highways. Dead squirrels littered the banks of the Hudson River, and an estimated 55 tons of drowned squirrels were removed from one New York reservoir.

The eastern gray squirrel is far more abundant in the relative sanctuary of suburbia today than it is in whatever forests remain in North America. In areas of extensive, suitable woodland habitat, grays occur in densities of between two and four animals per acre if the nut supply is good. In the wilderness the gray is a wily, furtive creature, quite unlike its bold and brash cousins of city parks and suburban streets. Among the homes and gardens of men, however, the squirrel is the single most abundant and noticeable small mammal throughout most of the East. On a recent fall bicycle jaunt through the picturesque, tree-lined streets of Spring Lake, New Jersey, I counted 19 squirrels occupying the environs of a single block in the town, nine of them boldly capering in one front yard alone. All the animals were fat and appeared in very good condition, indicating a food supply more than adequate to the support of such a large number of active, hungry rodents.

The suburbanized gray squirrel, living as it does in an environment free of human hunters and most natural predators, faces far fewer perils than its deep-woods brethren, but two constant threats—domestic cats and the family car—take a great toll every year. Uncounted thousands of squirrels die under the wheels of suburban traffic, but they are so prolific, and the food supply is so abundant and secure, that most populations remain stable.

The gray squirrel almost attained a sort of literary immortality in the beloved fairy tale *Cinderella*. According to the original version of the timeless fable, reportedly told in the Norman-French language, the word *vair*, meaning "squirrel fur," was used to describe Cinderella's famous slippers. Over the centuries and in countless retellings of the tale, the word became the similar-sounding *verre*, a French word meaning "glass."

Cottontail rabbit
Sylvilagus floridanus

The sight of a small brownish rabbit nonchalantly munching lawn grass or weeds beside a quiet suburban street or a busy superhighway is a little wildlife experience pretty much taken for granted by most people today. Although the cottontails as a group require some form of dense cover for protection from predation, they have shown an amazing ability to adapt to lawns and gardens, and can be found in or near larger cities wherever enough waste areas covered with edible weeds remain.

The cottontail's scientific name, *Sylvilagus*, combines the Latin *Silva*, meaning "forest," with the Greek *Lagos*, meaning "hare." Although many cottontails are indeed found in wooded areas, this animal by no means restricts its activities to such sylvan locales. Depending on the species, a cottontail might be spotted darting away in a dank marsh, a dry desert, an old-growth forest, a bucolic country lane, or over the weedy and trash-littered ground of an urban sanitary landfill.

Cottontails of one species or another exist over most of the United States and extreme southern Canada. The eastern cottontail (*S. floridanus*) is 14 to 19 inches long and weighs between 2½ and 3½ pounds. It has the largest range, occurring from central New England west to Minnesota and south to Florida and Mexico. The three other recognized species include the mountain cottontail (*S. nuttali*), the desert cottontail (*S. auduboni*), and the New England cottontail (*S. transitionalis*). The closely related swamp rabbit (*S. aquaticus*) and marsh rabbit (*S. palustris*) are somewhat smaller than the cottontail and, as the common names imply, prefer moister habitats. These cottontail species vary in color from pale grayish-buff in the desert cottontail to dark brown in the marsh rabbit. Regardless of color, all the various species conform to the overall cottontail description—a small, brownish rabbit with shorter legs than a hare and a white, powderpuff tail.

Cottontails, in particular the eastern species, have adapted well to the urbanized environment in part because they require and maintain smaller territories, seldom exceeding two or three acres. This modest territorial imperative eliminates the need for extensive wandering felt by larger mammals who must oversee their turf, and thus helps insulate the rabbit from the constant suburban threat of cats, dogs, and speeding cars. Though they are capable of digging their own burrows, cottontails prefer to use the unoccupied homes of other excavating animals such as woodchucks and foxes. For most of the day, the cottontail holes up in its "form," a shallow depression in thick vegetation that commands as wide a view as possible of the surrounding area.

This animal also depends on its admirable fecundity to renew its numbers. Two to four litters of between three and seven young are produced each season, depending on latitude. The unfurred and helpless babies are well hidden in a small depression in the ground lined with the mother's belly fur, and skillfully concealed from detection by a covering of grasses and other vegetation. The mother visits the nest only to nurse the young, and because they have little or no scent early in life, little cottontails are exposed to predators only when they begin to hop about in the vicinity of the nest after about two weeks.

Because cottontails eat a wide variety of vegetation, both wild and cultivated, they are regarded as garden pests wherever they are numerous. They freely eat all parts of the poison ivy plant without ill effect. In order to elude its many enemies, the cottontail can run at speeds up to 20 miles per hour, though it looks much faster due to the small size of the animal and the erratic, bounding nature of its flight. In spite of its swift skills in escaping would-be predators, about 85 percent of a given population die or are killed each year. A typical wild rabbit is lucky if it lives out its maximum lifespan of three to four years; they have been known to survive 10 years in captivity.

Millions of cottontails die on the roadways of the continent each year and many more are hunting casualties. Pennsylvania, a state having millions of these little rabbits, loses an estimated five million bunnies to predation, road kills, accidents and poaching in the period between September 1st and November 1st alone. The legal annual Pennsylvania hunting harvest alone averages between 25 and 40 million animals.

Woodchuck
Marmota monax

The roly-poly woodchuck is a marmot, a member of the Sciuridae, the large and diverse squirrel family. There are some 16 marmot species, most of them occupying mountainous habitats throughout Canada and much of the continental United Sates as well as western Europe and most of Asia. All marmots are active during the daytime, retiring to their often extensive burrows at night.

Also known as the groundhog and whistle pig, the woodchuck is one of America's best-known and most visible mammals, its reputation at least in part gained through the yearly media attention paid to Groundhog Day every February.

The woodchuck is a relatively large animal, reaching up to 26 inches in length, and weighs up to 20 pounds. An average adult 'chuck will weigh around 10 pounds. The species is widely distributed across the northern tier of the continent, from Labrador and Virginia in the east, to extreme eastern Alaska and southern British Columbia in the west. In much of the western United States it is replaced by the yellow-bellied marmot (*M. flaviventris*), a much less common or adaptable animal found in mountainous regions.

Although there is no way of tallying the number of North American 'chucks exactly, a 1980s survey based on suitable habitat came up with the figure of 500 million chucks continent-wide. The woodchuck's high rate of reproduction and its wariness wherever it coexists with people have helped this animal maintain its numbers despite intense hunting pressure (they're quite good-tasting), the perils of speeding cars, and longstanding predation by farm dogs.

The woodchuck can climb low trees and shrubs, and can run at a speed of about 11 mph, just about fast enough to outrun a farm dog if the safety of the home burrow is not too far away. With luck, they can survive up to five years in the wild, perhaps 12 in captivity.

The woodchuck is one of the true hibernators among mammals, and spends much of its life sleeping, winter or summer. During the summer months the animal can sleep 20 hours a day in its burrow, while in winter, its famed hibernatory sleep can last five months in more northerly latitudes.

During hibernation, the 'chuck's body temperature falls to about two degrees above the surrounding air temperature in the burrow. That two-degree difference shows that the 'chuck is not a cold-blooded animal, but a warm-blooded creature hovering just outside death's door in a period of almost complete cessation of bodily functions.

The den of this large rodent can be extensive, with any number of escape entrances. There might be between 5 and 19 "back doors" in the maze of burrows, most of the escape hatches concealed by rock piles or vegetation. The burrows are often shared by cottontails and red foxes, especially during the winter months when the 'chuck is in hibernation.

As wild animal pets go, the woodchuck makes a good one if it is captured when very young, within about two weeks of leaving the natal den. Most 'chucks become quite tame and affectionate if treated gently and handled frequently, though the majority of them, especially the males, can become nasty and unpredictable as they age. The animal has the typical gnawing teeth of the rodent, but a large 'chuck will defend itself with determination and ferocity and can deliver a dangerous bite.

This rather large and conspicuous, day-active animal is able to persist at the very edges of cities and in suburban areas due to the lack of hunting pressure and the absence of natural predators. Human-created environments such as golf courses, corporate parks, airport verges, and highway shoulders offer an endless supply of lush, easily-grazed summer greenery. As long as the wary 'chuck can avoid the occasional stray dog, the traps of groundskeepers, and the crossing of busy highways, it seems assured a place in the urbanized environments of tomorrow.

Coyote
Canis latrans

The coyote, or prairie wolf, might be the truest testament to animal survivorship in the face of relentless human persecution— against the longest survival odds imaginable. For the 10,000-odd years that man-the-hunter has engaged in competition with the "prairie wolf" for available game, he has tried to eliminate it from the natural scene—a campaign that has still not met with success though it continues with determination in most parts of the West. In 1991, the Federal Animal Damage Control program, essentially a taxpayer-financed extermination service for ranchers, killed 2.5 million animals, the bulk of them coyotes.

The coyote's common name is derived from the Indian name coyotl. The scientific name, Canis latrans, means "barking dog," something the coyote seldom does. Rather, it "sings" with the wonderful medley of yaps, howls, whines, and other canine vocalizations so characteristic of the Old West, though increasingly heard in the mountains of New England.

The original range of the little prairie wolf is something of a mystery, for it might well have occupied most of the continent prior to the coming of Europeans, though no doubt the animal kept well out of the way of the gray wolf, which then inhabited nearly all of North America. In historic times, however, the coyote was primarily a western animal, ranging from central Alaska and the Northwest Territories west to the Great Lakes, and south into Mexico and Central America. Since the virtual extermination of the wolf over much of North America at the end of the last century, the coyote has extended its range eastward, occupying the much larger carnivore's former haunts with alacrity. The coyote has in recent years been observed and trapped in virtually all of the eastern states, where several subspecies have been described. It is particularly numerous throughout northern New England, where so-called "coy-dogs" have been melodious fixtures in the Green and White Mountains for years. The animal was first reported in New York State in 1912, and it is now common in the Adirondacks. In eastern Canada, coyotes moving into regions occupied by relict populations of timber wolves successfully interbred with these larger canids, and the result is a race of larger coyotes with thicker coats better suited to the damp and colder East.

Coyote

Although the coyote can be found in nearly every conceivable type of natural habitat, and is even spotted in urban situations, it much prefers open areas where it can keep an eye out for distant threats and get a good head start if they materialize.

In appearance the coyote suggests nothing more or less than a small dog with a bushy tail. Unlike the typical dog, the tail is held downward, usually tucked between the legs, when the animal runs. The color ranges from a dull yellowish gray to darkish-gray and rust. The legs, feet, and ears are usually a rusty red, and the belly is whitish. The coyote's face and head are rather lean and foxlike, the golden eyes very bright and expressive. Its overall appearance suggests the high degree of intelligence for which it is famed. The average adult weight is 18 to 30 pounds, with the largest recorded a 75 pound male; the female is about one-fifth smaller than her mate.

Smaller than the wolf, and less committed to a tight social organization, why has the coyote persevered while its larger and more powerful cousin is clearly on the way to extinction? The coyote's continued survival in the face of unrelenting human persecution and a drastically altered natural environment is a result of the animal's physical and temperamental attributes. The coyote's very behavioral plasticity and willingness to eat anything edible has doubtless contributed to the animal's survival success.

The coyote trots at 10 to 15 mph, gallops at 25, and can reach a top speed of about 35 mph if it has to. If it manages to evade all of its many human and natural enemies, the life span of the coyote can, with luck, reach 8 to 10 years in the wild. The animal has survived up to 18 years in captivity, about the same longevity enjoyed by the average, well-fed, medium-sized domestic dog.

Red fox
Vulpes vulpes

The canny and adaptable red fox has one of the broadest ranges of any North American mammal, occurring from arctic Alaska and eastern Canada south to California, New Mexico, Texas, and the Carolinas. Once known as *Vulpes fulva*, this fox is now considered to be the same species as *V. vulpes*, the red fox of Europe.

This little canine is primarily associated with rural and agricultural locations, though it can persist among the subdivisions of the new suburban environment as long as intense urbanization does not remove its food supply. Living up to its nickname, "Reynard," or the "trickster" (a name applied to the monkey, the raven, or the coyote, according to the culture), a fox can avoid conflict with dogs and people as long as familiar escape routes are available, and plenty of den sites and protective cover are present in its territory. As with all smaller mammals, a fox cannot cope with speeding cars. Although suburban red foxes are quite "street smart," many are killed on our increasingly busy highways each year. Under normal conditions, a red fox can survive up to 10 years in the wild; captive and pet foxes have lived 15 or more years under good care.

The color of the red fox can vary from light blond to the familiar deep russet-red. The snout, ears, and paws are normally black, and the throat and belly a clear white—a very beautiful and classy creature! Other color variations include the "bastard fox," a red fox with a dark, bluish-gray coat; the "cross fox," with a cross-shaped patch of darker hair on the shoulders; and the so-called "Sampson fox," which lacks the long outer guard hairs, having only the buffy or brownish woolly undercoat. The name stems from the Old Testament account of Sampson's rout of the Philistines by tying burning firebrands to the tails of foxes and setting them loose in the enemy's crop fields.

The red fox mates in January or February throughout most of the range, and the animals pair for life unless disturbed by human activity. Three to ten kits are born in April in the often extensive burrow; within about two months they follow their parents on hunting trips. By the time they are six months old they are on their own and capable of carefully assessing carrion for the hidden danger of traps.

The red fox, in both the pup and adult, loves to roll in malodorous materials, such as dead animals and feces, just as domestic dogs do. The cause of this odd behavior is unknown, though it no doubt has something to do with the expression of territoriality and canine identity.

Although the number one food of the red fox is the meadow vole, with the cottontail rabbit a close second, the animal is not an overly specialized carnivore and will eat a wide variety of vegetable foods. As suggested by the fable "The Fox and the Grapes," the red fox has a weakness for grapes, cherries, blueberries, and rose hips, as well as many items of garden produce. The animal is thus much less dependent on the abundance of animal prey than other carnivores, and can readily exploit the biologically impoverished environs of the agricultural and urbanized landscape.

The red fox's smaller size and ability to remain hidden even in densely settled areas will doubtless serve it well in the human-dominated world of the future. Agile and catlike in behavior, foxes hunt alone or in well-coordinated pairs, attracting less attention than the pack-hunting tactics of the larger wild canids.

Raccoon
Procyon lotor

"The Chief Beasts of Virginia are Beares, lesse than those of other places, Deere like ours, Aroughcun much like a Badger but living on trees like a Squirrell." With those words, by an early 17th century American colonist, was the adaptable and familiar raccoon first introduced to the annals of the natural history of the New World. Over the ensuing three centuries the spunky animal's Indian name underwent gradual contraction, to arakun and then raccoon, or simply 'coon.

The raccoon is a procyonid ("before dogs") mammal related to the ringtail "cat", coati, and the bears. It is a plantigrade animal; that is, it walks on the flat, heel-to-toes area of the foot, much in the manner of bears and people, as opposed to the digitigrade, or toe-walking locomotion employed by dogs and cats. The raccoon's full scientific name means, in essence, "the before-dog who washes his food," and pretty much sums up this adaptable, vaguely doglike creature's image in the eyes of humans, who find themselves sharing more and more of the 'coon's habitats.

The raccoon is found over most of temperate North America, from southern Canada south into Central America. It is absent only in parts of the Rocky Mountains and some extremely arid areas of the western deserts. For the most part raccoons have loosely defined territories that might encompass some 300 to 400 acres. They travel irregularly over rather casually defined routes, usually by night, and by dawn are safely back in a permanent den. This might be located among rocks, in a large dead tree, or increasingly, in an unused house chimney, or beneath a building. When the young of the year disperse, they will often wander great distances—up to 130 miles—in search of territories and dens of their own.

The combination of an abundant food supply and the protection offered by man's structures has often pushed local raccoon populations to abnormally high densities in suburbia. In a suburban area just outside Cleveland there were found to be an average of one raccoon per 1.4 acres, as opposed to one animal per twelve to forty-five acres in rural areas and wildlife refuges.

The raccoon is one of the larger mammal survivors. An average adult raccoon will weigh 12 to 16 pounds, but in spite of its size and bulk it is an extraordinarily flexible creature. The average 'coon can squeeze through an opening a mere 3½ by 4 inches big, which helps to explain why they are able to exploit such a wide range of urban situations for food! Jumbo 'coons weighing 40 pounds or more are not uncommon in suburban areas today, and the largest ever reported was a veritable giant of 59 pounds.

Raccoons are opportunistic omnivores. They will eagerly devour anything and everything edible encountered in their nightly travels, from fishes and frogs to the eggs and young of birds, and all kinds of organic garbage and carrion. In marshy areas they can play havoc with the resident muskrats by locating the rodents' grass lodges, tearing them open, and devouring the young muskrats found within.

Where in the past the raccoon represented a rather mildly troublesome holdover of the wilderness in suburbia, today the advent and spread of rabies among wild raccoons throughout the urbanized Northeast has given the animal's relationship with mankind a more ominous overtone. Over the past five decades the raccoon has been a regrettably familiar part of the roadkill "fauna" of the eastern United States, but in the past five years the numbers of traffic-flattened 'coons has greatly increased, especially in the rapidly expanding suburbs. The principal reason for the carnage is the large number of rabid and thus disoriented raccoons wobbling across roads in broad daylight, and otherwise coming into unnatural contact with man, his machines, and his domestic pets. A nocturnal creature, the adaptable 'coon has long been able to coexist with man by avoiding confrontation. But a sick and bewildered raccoon wandering about by day is a large, conspicuous, and potentially dangerous animal that quickly attracts the attention of neighborhood canines, or is hit by cars at peak traffic times of the day.

The current rabies epidemic among raccoons and several other wild mammal species apparently had its origin in an undetermined number of carrier animals transplanted into West Virginia as replenishment of hunting stock. These raccoons reportedly were exported from Florida and several areas of the Southwest in the mid-1960s, where the disease was present but certainly not in epidemic proportions. The first rabid raccoons in the Middle Atlantic region were reported in West Virginia in 1966, and from there the disease progressed northward, initially at a slow rate. Rabies among wild 'coons appeared in neighboring Virginia in 1978 and in Maryland three years later. By 1982 rabid animals were reported and collected for analysis in Pennsylvania, and in 1987 a number were shot in Delaware. Over a period of three years—from 1989 to 1991, the disease appeared in three heavily populated states. New Jersey, New York, and Connecticut, respectively, all reported rabies among their wild raccoons. The more mountainous, less populated states of northern New England are expected to feel the effects of the epidemic within the next few years as the disease continues its northward expansion.

As the disease has already turned up in wild animals in both Florida and North Carolina, and there have been isolated reports in Ohio and New Hampshire, those states can be said to be the present geographic limits of raccoon rabies—for the present.

Raccoons are amazingly abundant, even in urban and suburban areas, and control of the disease by removing most or all of the animals is a virtual impossibility. The next best thing might be a new oral vaccine that has been successfully field tested in Pennsylvania and Virginia, and at this writing was due to be tried in New Jersey. The vaccine, reported to be effective in reducing the incidence of rabies in wild animals, is applied through aerial spraying of severely affected areas or by treating special baits or other foods that foraging raccoons eat.

The three commonsense steps the average person can take in the face of the raccoon rabies threat are:

- If you have a dog or cat, make sure it is vaccinated against rabies.
- If you are bitten by any animal, even a household pet, seek medical treatment immediately. Make an effort either to positively identify the animal or bring it, alive or dead, to health officials so that it can be tested for the disease.
- Avoid close contact with any wild animal, especially one that seems "tame" or is behaving in a strange or unnatural manner. In other words, never attempt to approach and pet a "cute little 'coon" seen wandering about in your backyard on a bright sunny day!

Opossum
Didelphis virginiana

The opossum, more correctly known as the Virginia opossum, but more popularly as just 'possum, is one of the more widespread—though unobtrusive—mammals among us today. Essentially a creature of warmer southern latitudes, the night-loving opossum has been steadily expanding its range northward since the 1890s, and is now found as far north as extreme southern Canada. The closely related and similar Mexican opossum (*D. marsupialis*) occurs from northern Mexico throughout much of Central America. The Virginia opossum has been introduced on the West Coast and now occurs from California north to southern British Columbia.

Though an exceptionally hardy creature, the 'possum cannot tolerate extended periods of subfreezing weather. It is not a true hibernator, and thus is forced to remain active year-round. It can survive harsh winter weather, however, as long as food, forage, and shelter is available and 'possums missing ears and tails due to frostbite have been regularly observed in New England. The looming possibility of global warming might, at least in the case of the opossum, have a beneficial, range-extending effect.

Surprisingly for such an easy-going, unflappable creature, the average natural lifespan of the opossum is rather short—a mere three years in the wild state at best—though captive individuals and those observed over time in more secure suburban backyard situations have survived for over seven years. In the northern parts of the range opossums depend heavily upon road carrion, household garbage, and the largess of suburban wildlife lovers for the sustenance required for winter survival.

The near-sighted, slow-moving animal is virtually defenseless against modern traffic. The automobile is probably the opossums' greatest enemy in the world of man. Millions of 'possums die annually on the roads of North America, and in many parts of the United States it is the most commonly seen roadkill.

Few carnivores, including domestic dogs, will kill an opossum with the intention of eating it, and its famous "play dead" reaction to an attack has doubtless saved many a possum's life in a pasture or backyard encounter with a family pooch. The great horned owl, a large, powerful, and primarily nocturnal avian predator that might remain abundant in suburban areas, includes both the opossum and the skunk in its diet. Prowling house cats—even large, street-wise toms—are no match for the muscle-power and sharp teeth of an adult opossum, and generally avoid any conflict with the animal.

Nine-banded armadillo
Dasypus novemcinctus

This odd little creature, seemingly an anachronistic, nearsighted holdover from prehistoric times, has made some pretty amazing gains over the past 100 years. The armadillo is an environmental success story. It did not arrive in North America by accidental or deliberate human agency, nor has it attained pest status along the way. It simply crossed the border, quietly and unheralded, from its native South America and Mexico into Texas less than 150 years ago, and has since bumbled and snuffled its way over the landscape and into the folklore and the menu of the Southwest. Since invading Texas, in the area between Brownsville and Rio Grande City, the armadillo has steadily extended its range eastward and northward, so that today the creature is a common resident from Florida and southern Georgia west to Arkansas and Oklahoma. It has long been established in Louisiana and Alabama, and in Florida the creature is rare or absent only in the extreme southern and central parts of the state. Recent scattered reports indicate that the animal is well on the way to establishing itself in New Mexico, Kansas, and Missouri, and a single gravid female was reportedly taken in Colorado. Armadillos were first reported in Tennessee in 1974, and there are numerous South Carolina records dating from the 1960s.

Armadillo

Although the armadillo appears to be intent upon conquest of the north like that other Southerner, the opossum, there are several mitigating factors that will ultimately limit its range expansion. Unlike the opossum, the armadillo cannot tolerate prolonged cold spells, and thus is unlikely to colonize the central and northern states, at least in the near future and under present climactic conditions. Under controlled conditions, armadillos died when exposed to temperatures in the 40s and 50s (Fahrenheit) over a two-day period. In addition, the animal requires a reliable, year-round source of surface water and succulent foods. This critical factor has prevented it from expanding its range into desert or arid brushy areas; it usually suffers declines in regions experiencing extended periods of drought.

The armadillo is an edentate, or animal "without teeth," though it does possess primitive, peglike dentures. The group includes anteaters, tree sloths, and approximately nine genera and 20 species of armadillos. The name *armadillo* is Spanish for "little armored one," referring to the flexible but tough, platelike shell covering the creature's body, unique to armadillos.

The nine-banded, or *novemcinctus*, armadillo shares with all the members of the family a unique, vaguely reptilian appearance the tanklike shell confers upon the animal. The shell consists of three main parts: a scapular shield over the shoulders; a pelvic section; and eight to eleven (usually nine) bands of a central section that joins these shields together. The top of the head is heavily scaled and the long, muscular tail is protected by about 13 leathery ringlike plates. Unlike some of the smaller, tropical armadillo species, the nine-banded cannot completely fold itself up into its shell for protection; instead it relies on its rapid running speed and a phenomenal ability to dig itself out of sight in a flash. On open ground a human in good shape can catch a fleeing armadillo, but in broken or wooded country the animal can easily lose all but the most agile and determined of pursuers.

The armadillo is far from gaudy, ranging from a dark brown to pale grayish-tan. An adult can be 2 feet long and weigh up to 14 pounds; the male is slightly larger and heavier than the female. The animal is a completely harmless insectivore lacking canine teeth or incisors. It might on occasion vary the bug diet with earthworms, small reptiles, amphibians, and young birds or eggs if the opportunity presents itself. They reportedly will kill and eat snakes, and have been introduced into parts of Texas specifically for that purpose.

In turn, armadillos have served as food for people for just about as long as the two have coexisted in the Southwest. The animal's reputed porklike flavor earned it the nickname "poor man's pig" during the Great Depression, and although its meat has yet to find a ready market outside its range, armadillo chili and other 'dillo dishes are widely enjoyed from New Orleans to Nogales.

The one factor that prevents this animal from finding favor with more human gourmets is the armadillo's high degree of susceptibility to Hanson's Disease, or leprosy. Leprosy, caused by the bacteria Mycobacterium leprae affects the skin, nerves, and some organs, and is greatly feared among humans for the physical deformities it causes. No one knows for sure how the disease got started among armadillos, but some researchers suspect that the animals picked it up from bandages discarded by Central and South American leprosy patients in the days before the widespread use of sulfones to treat the illness.

Since the armadillo is the only known animal species that regularly develops and carries lepromatous leprosy, it has been used extensively in the search for a vaccine for the disease. In tests, 80 percent of the armadillos exposed to the disease contracted it, and fears have been raised that the illness can be transmitted to man through the handling of infected animals. While it is agreed that there is at least some risk in handling or eating armadillos, contracting this only moderately communicable disease would require the handling of many live animals or their raw meat and tissues over several years' time, and sustaining cuts or abrasions in their capture or cleaning.

The shells are made into baskets and lampshades, and the meat has been extensively canned for human consumption. The armadillo was nominated, but subsequently rejected, for the honor of state mammal of Texas.

For the most part, the armadillo's future looks fairly bright. The creature seems well able to adapt to the spreading and well-watered subdivisions and golf courses of the south central and southeastern range core-area, and can probably continue to thrive under increased urbanization as long as it can avoid speeding vehicles, against which neither the armadillo nor any other small wildlife has any real defense. In addition, given the current human fascination for the armor-plated little critters and the spate of armadillo races sponsored each year by the World Armadillo Breeders and Racers Association, the animal is almost certain to be transported by 'dillo lovers to unconquered areas where they might well escape their captors or be liberated to establish themselves anew like many of the wildlife survivors of the past. Indeed, researchers have found that the release of a single pregnant female armadillo, or even a pair of the animals, is sufficient to start a new resident population.

Striped and spotted skunks
Mephitis mephitis and Spilogale putorius

Striped skunk

Spotted skunk

Skunks, or "wood pussies," are first-class exploiters of the urbanized environment, being omnivores of catholic food preferences and nocturnal habits, and carrying chemical weaponry that few other creatures are inclined to challenge. Skunks share their musk-producing capabilities with their other relatives in the Mustelidae—the weasels, ferrets, and the formidable wolverine of the northern boreal forests.

The spotted and striped species are the two most widespread and abundant North American skunks. The spotted skunk occurs from Central America north to Washington State, Wisconsin, and Maryland, and in the East primarily following the Appalachian Mountains. It is absent in New England and along the Atlantic Coastal Plain. The larger striped skunk, more familiar to most suburban residents, is found from central Mexico to northern Canada and east to Labrador and Nova Scotia, and rare or absent only in extreme southern Florida.

The spotted skunk attains a length of about 20 inches, including the tail, and weighs between one and two pounds. Individual skunks are generally smaller in the western parts of the range and larger in the East. This is a handsome little creature, glossy black and sporting bright white spots on the forehead and under each ear, as well as four broken white stripes along the sides. The pattern is highly variable among individuals, and the tail might or might not have a white tip.

The striped skunk is a bigger and heftier animal, reaching a length of about 28 inches including tail, and weights of up to 10 pounds. Its familiar appearance scarcely requires a description. In general, it is a black, house-cat-sized animal that has a white stripe on the forehead, and a broad white area on the nape that usually divides into a V on the shoulders. As in the spotted skunk, the pattern is highly variable and mostly white and mostly black skunks are often seen. Its pointed snout and beady eyes reflect its weasel relationship, though with its flat-footed, shuffling gait and rather easy-going demeanor, it is worlds apart from the high-strung, insatiably carnivorous little mustelids that share the family with it.

The nocturnal presence of a skunk is often detected by odor rather than by sight, and its unique defense is perhaps its greatest hallmark. The skunk's scent glands consist of two oval sacs located just beneath the skin under the tail. The animal can spray (perhaps "shoot" would be a better word) its pungent musk via twin ducts with some force and can hit targets up to 10 to 15 feet away. When in use the scent ducts protrude from the skunk's body so that the animal's fur is not drenched and permeated with the owner's odorous and long-lasting perfume. The skunk must execute an about-face and point its rear end toward a persecutor before it can fire with any accuracy, so a skunk facing a curious human poses no immediate threat.

The skunk's spray can cause excruciating pain and temporary blindness if the eyes of an attacker sustain a direct hit. The incredibly persistent odor can be removed by repeated scrubbings with acidic tomato juice, though the olfactory remembrance of an attack can linger in clothing or in a dog's coat for up to year.

Both the spotted and striped skunks live in a wide variety of habitats, and both adapt well to the suburban scene, often utilizing houses, both abandoned and occupied, commercial buildings, and other constructions as den sites. As they are almost entirely nocturnal in behavior, their presence can go undetected by most human residents of an area unless the skunk has an unfortunate encounter with a family dog or a passing car. Roads and the vehicles that travel them at today's high speeds are probably the skunk's greatest enemy throughout the more highly developed parts of its range. It is well known that the odor of a roadkilled skunk can waft up and down a highway for a mile or more in either direction, particularly on a damp night.

As opportunistic omnivores, skunks compete directly with raccoons for the abundant provender offered by the garbage cans, dumps, and roadsides of the urban and suburban scene. But as skunks are rather placid, peaceable creatures by nature, the two species coexist with relative harmony. A raccoon, unlike an overzealous domestic dog, knows only too well the potency of the skunk's chemical defense, and keeps its distance under most circumstances. Even when confronted by a human, a skunk is likely to simply pause in its doings and peer nearsightedly at the intruder with mild curiosity rather than take any threatening action of its own—unless the other institutes hostilities. In general, these animals adhere to a "live and let live" philosophy.

Skunks will eat just about anything organic and edible; insects, worms, fruits, berries, roadside carrion, and garbage are all welcome fare, and striped skunks have been observed rolling toads and large caterpillars in grass to remove their toxic exudates before eating them. Where they are abundant these animals can pose a considerable threat to ground-nesting birds and poultry.

White-tailed deer
Odocoileus virginianus

The white-tailed deer is a true North American native, having evolved here over 10 to 20 million years ago in the Pliocene and Miocene eras. The deer as we know it today took form about a million years ago in the Pleistocene era.

By the 1890s the whitetail was in dire straits due to the virtually unrestricted hunting it had endured since colonial times, having suffered near-catastrophic declines throughout its range. New Jersey had less than 100 deer at that time, and the states of Connecticut and Rhode Island had none at all.

The story of the decline and resurgence of the whitetail can be quite dramatically told using the figures of my home state, New Jersey. In 1900, following many years of relentless persecution, the state's deer population was estimated at less than 100 individuals; today, the herd is reported as numbering somewhere between 140,000 and 160,000, due to the prohibition of uncontrolled market hunting, effective state management, and the nature of the prevailing mixed suburban-agricultural habitat. The 1992 summer census revealed a population of some 146,000 animals, with a projected 1992–93 harvest of 47,000; the 1989–90 deer harvest, considered a record, was 48,000 animals. Road kills have accounted for between 8000 and 10,000 animals yearly since the 1980s.

When absorbing statistics like these, keep in mind that I'm not talking about big, essentially rural states like Maine or Montana, but New Jersey, the most heavily urbanized and densely populated state in the Union. In New Jersey, the deer has about three million acres of at least seminatural land available to it. Although much of this area is relatively undisturbed woodland habitat, a large part of the territory is "edge" habitat, rapidly suburbanizing, and the animals' presence there can be very much a mixed blessing to the human inhabitants. New Jersey wildlife management officials feel that the state's deer harvest might have reached a peak in the 1992-93 season never to be attained again, as spreading development consumes more and more deer habitat.

Much of the current deer problems in the state, namely those involving too many deer and their unwelcome contacts with too many people, stem from the patchwork nature of the habitat. Sizeable parcels of deer habitat are state or county parks, and thus off limits to the state's 165,000 licensed deer hunters. In addition, other tracts might be privately owned and thus posted against hunting, allowing the resident deer to reproduce themselves. Given an adequate diet and good habitat, a doe can reach sexual maturity in a year, and can bear up to three fawns at a time, though two is the norm. In the absence of predation, whether from wild carnivores or human hunters, the deer population of a large wooded tract surrounded by suburbs can quickly reach critical mass. The side effects are then deer-people encounters of the worst kind: dangerous road accidents, pilfered gardens and flower beds, and attacks by roving dogs. New Jersey game officials say that without hunting pressure, the state's herd would eventually increase to around 500,000 animals; the accompanying stresses of its coexistence with the huge human population can only be imagined.

Deer generally live in a matriarchal society; that is, outside of the mating season, the herd is led by an older, experienced doe, much in the manner of elephants. They begin feeding at about 4:30 pm and filling up usually takes about an hour, after which the animals retire to a sheltered, quiet place in which to chew the cud. A healthy adult deer requires some 10 to 20 pounds of fodder daily to maintain condition.

White-tailed deer

The principal myth concerning the whitetail is that of its size; people think the deer is a lot larger than it actually is. An adult buck will stand between 36 and 40 inches at the shoulder and weigh around 150 pounds—not much bigger than one of the larger dog breeds like the Rottweiler. However, the largest reported buck weighed a gargantuan 425 pounds.

The lifespan of the whitetail under natural conditions is not surprisingly short; the average is about nine years, with few bucks—about 2 percent of them—even making it to their five-year prime before they fall victim to a hunter's bullet. Captive deer can live well into their teens and some have reached the 20-year mark, but this is exceptional.

That the deer is a speedster in the race for survival goes without saying; its reliably recorded top speed is between 30 and 35 miles per hour. But given the nature of the country the deer inhabits, the ability to negotiate stands of dense timber and tangles of shrubbery in a flash more than outright speed is usually the key to eluding pursuers. A healthy adult deer can clear an eight-foot fence without trouble, and can broad jump 15 feet. A record jump of 28 feet was reported.

The ultimate legacy of the whitetail's status as a mammal survivor can be found in its reputation as a pest species throughout much of its range. Where the deer coexists with mankind and his gardens there is little love lost and the animal, munching its gracefully innocent but insatiable way through the tomatoes and hydrangeas, is caught solidly between the attitudes of their human admirers who would prohibit the killing of even one for any reason (the "Bambi Syndrome" of the hunting fraternity), and the beleaguered suburbanites and truck gardeners who see something less than an endearing Bambi in the sight of a regal deer among the bean sprouts. In between are the wildlife management people, who admire the deer for its grace, its beauty, and its place in nature, but recognize the need to limit its numbers in this increasingly urban age.

As a result of the growing human and whitetail conflict, many states undertake periodic culls of problem herds—programs that are always unpopular with the general public—and have begun experimenting with immunocontraception as a means of limiting deer numbers. In the latter program, females are shot with a dart carrying a protein vaccine that blocks fertilization. This approach to the deer problem has been tried with some success in Virginia and California and is scheduled to be implemented on New York's Fire Island in the near future.

Large-scale farmers often suffer large crop losses to deer, but suburban gardeners can have their produce and the deer as well if a few precautions are taken. Deer can be fenced out of gardens, but the height of the barrier required to do the job will vary from 8 to 11 feet, depending on who you listen to. There is no doubt that a healthy adult whitetail can easily clear a five- or six-foot fence if the produce on the other side offers enticement enough, and they have been known to crawl beneath carelessly installed fencing that offers any sort of ground-level access.

Aluminum pie pans hung from wires and "scaredeers" set up on the perimeters of gardens sometimes work, though on windless days and nights, people-wise deer will graze beside them with impunity. A dog tied near a garden will serve as a true deterrent, with the only real drawback being the need to post the unfortunate pet there day and night for the duration of the growing season.

Aside from shooting them, the only alternative to fencing deer out of the family garden lies in determining what the animals won't eat and either featuring these plants as ornamentals or growing them in the form of a living barrier. In general, deer are put off by tough, hardy, aromatic plants, those that give off an almost oily odor. Chief among these are lantana, yarrow, caryopteris, nepeta, rosa rugosa, and bayberry. Foxglove, globe thistle, gaillardia, and salvia are also proven deer deterrents, along with nearly all evergreens except yew, which the animals find very attractive. Ordinary privet hedge is usually safe, though while this common ornamental shrub might go untouched in one area, the deer of another district might chomp it right down to the ground.

The most effective approach to deer discouragement varies from region to region, and thus what works on New York's Fire Island, where there are about 600 nearly tame deer ravaging the summer residents' produce, might not do at all in the mountains of Virginia. For a general, commonsense overview of the deer-garden problem and its solutions the reader is referred to an interesting little book entitled *Gardening in Deer Country*, by Karen Jescavage-Bernard. The paperback sells for $7.10 and can be had by writing the author at 529 East Quaker Bridge Road, Croton-on-Hudson, NY 10529.

Outdoor writer Pete McLain perhaps sums up the image and the enigma of the modern whitetail when he says that "the present-day wild deer is looked upon by most of the public as something wild and beautiful in the woods. Farmers who suffer deer depredation on their crops see the animal as a four-legged nuisance they could live without. Deer hunters see the deer in terms of outdoor recreation and a supply of high quality meat in the freezer. The animal rights people view the deer as a creature that must be preserved at whatever cost."

Chapter eight

The roadkill factor

Many men now living can remember the coming of the bicycle, the bicycle followed by the automobile and the coming of hard surfaced roads, pavement laid down in small town streets and sewer systems installed. The paved roads crept out in all directions, found their way into isolated coal mining towns, into the hill country that in the East and Midwest separates the North from the South, reached out across the Great Plains, the mountains and the deserts, to the far West.

Sherwood Anderson, _Hometown_ (Alliance Books, 1940)

The wheel and axle came into being at least 5000 years ago. According to the archaeological record, wheeled vehicles originated in the region south of the Caucasus Mountains, and in eastern Turkey and northern Persia. They were plying the roads, such as they might have been at the time, sometime prior to 3000 B.C.

In contrast, early American civilizations never produced wheeled conveyances and thus the early roads—really benign footpaths—on this continent were designed primarily for pedestrian travel or at best the riding of beasts of burden. Europeans arriving here in the 17th century found no ancient

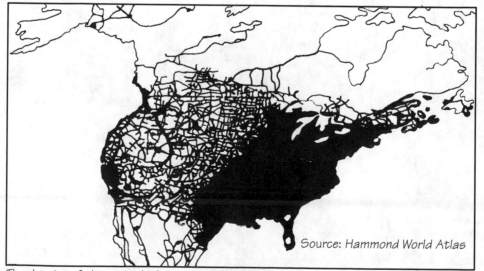

Source: Hammond World Atlas

The density of the major highways in North America

roadways, but rather an extensive network of forest trails used by the native Americans for travel and commerce. These precolonial arteries were environmentally friendly by today's standards, and took little or no toll of the landscape and wildlife. In 1707, the first regular freight service road was built in New Jersey. This roadway, running from Perth Amboy southwest to Burlington to link the small cities of New York and Philadelphia, is considered the forerunner of the modern road, at least in the New World.

The early roads had little direct effect on the native flora and fauna of this continent until the early years of the current century. But then things began to change, and it took the invention and development of the automobile to usher in the time of trial for wildlife.

The continental United States encompasses some 3.6 million square miles of land surface and at present has a little more than 3.6 million miles of roads. The total area of roads in the U.S. covers about 24,000 square miles, an expanse equal to the area of the state of West Virginia. In all, American streets and roads occupy about 22.1 million acres, or about 1.16 percent of the total area of the continental United States. That figure, however, is steadily growing; even as you read these words, new streets and roads are bulldozed into undeveloped land to create new subdivisions, and vast parking lots are cleared and paved around new shopping malls and industrial complexes.

The interstate highway system presently encompasses more than 44,000 miles of multilane roadway and as each mile of expressway pavement alone consumes, on average, about sixteen acres of land, the extent of that federal project's impact on the living landscape can be imagined.

Urban acreage is increasing by about 400,000 acres annually. Because about 30 percent of the urban landscape is devoted to roadways, some 120,000 acres of the spreading megalopolis are added to the urban and suburban road network each year.

The invention and development of the internal combustion engine and subsequently, the automobile, is an historical event that has had a profound and permanent effect upon the flora and fauna of virtually every landmass in the world. In the United States, the private car and the millions of miles of roads that accommodate it have shaped the landscape of a continent and determined its boundaries and parameters as no other force, natural or manmade, has been able to do since the onset of the last glaciation 12,000 years ago.

It is difficult today to experience a view of the American landscape without the intrusion of an automobile or the highways it travels. It's nearly impossible to find a wilderness spot in the lower 48 states that is five miles or more from a road of some kind, whether a two-track logging road or an interstate freeway. Access to the automobile has been granted virtually everywhere a road can be laid, even over the highest mountain ranges, and over or under the broadest bays. The four-wheel-drive and all-terrain "adult toys" snatch the baton from the street machines where the pavement ends, probing and invading every last bastion of nature wherever it might be found, from the high desert to the sea level swards of the coastal marshes. Every living thing on this continent, and throughout the remainder of the world, has been, is, or inevitably will be affected by the automobile— often violently so—along with its numerous satellite and parasitic accessories.

The United States today is swarming with some 251 million private cars and trucks—about one for every living person—and that number shows no signs of declining as ever more fledgling drivers come of age and join the mobile throngs already on the road. America has had a long and ardent love affair with the private automobile, and convincing the vast majority of citizens that it would be in the environment's best interests to phase out the car and resurrect the nation's ailing public transportation system will be far from easy.

The situation worldwide is even more disturbing and ominous, not only from the standpoint of highway building, but also the growing specter of air pollution and acid rain, both generated in large part by road traffic. At the beginning of the 1980s there were an estimated 310 million passenger cars alone worldwide. Today that number has grown to a little more than 400 million and by the end of the century there should be some 537 million passenger cars out of nearly 680 million motor vehicles of all types plying the highways of every continent except Antarctica. If present trends continue, by the year 2020 there will be double the number of vehicles on the world's roads and nearly double the number of miles driven per capita as are driven today. The great majority of these vehicles, particularly those rolling wrecks chugging and backfiring over the roads of many Third World nations, have nowhere near the pollution controls that are now mandated in North America.

Car ownership has actually peaked in the U.S. as statistics show that in 1992, there were 0.95 cars per person of driving age—just about one car for every man, woman, and child over the age of 16. In October, 1992, as the nation was emerging from the grip of the recent recession, automakers found hope in the report that some 10 million new cars were produced domestically and would soon add to the throng already clogging the roadways of North America.

In this country the average car is driven 9900 miles per year. For the period between 1965 and 1987, the average miles driven per person increased by 69 percent. That average increased by a whopping 154 percent in Europe, a continent that has seen a tremendous surge in private car ownership over the past two decades. Perhaps the most ominous aspect of this powerful desire for personal mobility, as well as the perceived status offered by ownership of the private car, has been the steady decline in the development and maintenance of public transportation. The car is king virtually everywhere people can afford to buy one and keep it in gas, and the network of habitat-destroying and fragmenting traffic arteries continues to expand like a malevolent cancer upon the living landscape.

The broad environmental effects of the automobile and the roads, parking lots, and junk yards that are its inevitable camp followers have been well documented. But what of the direct impact, literally speaking, upon animal populations and individual creatures? In other words, what about the role of the roadkill factor in animal populations' decline and survival?

Published research in the area of the annual roadkill toll and its effect upon animal populations and movements is rather sparse, but the figure of at least 400 million reptiles, birds, and mammals perishing yearly on the nation's highways is often cited. This figure is an educated guess based on the reasonable assumption that, depending on the time of year, up to one million animals perish on the nation's roads on any given day; the actual figure is probably much higher, taking into account those wild creatures that are not killed outright, but crawl off to die later.

Author Roger Knutson, in his engaging little book *Flattened Fauna*, cites historical estimates that place the number of roadkilled animals at "from 0.429 to 4.10 animals per mile of prime highway habitat." Knutson's book addresses the roadkill dilemma with a sort of resigned, tongue-in-cheek humor, but he does offer some disturbing statistics. According to the "logbook of a dead-animal aficionado in California," nine reptiles, 58 birds, and 161 mammals were observed dead-on-the-road during one 480-mile trip in 1984. Another, much older tally reported "598 rabbits on 50 miles of two-lane asphalt road near Boise, Idaho" during a motor trip undertaken in 1933.

There is sparse published material on the roadkill factor prior to the 1930s, however. One Iowa researcher who did take the trouble to comprehensively study the roadkill problem in that state in 1938 came up with the figure of 0.426 dead organisms per mile of high-speed roadway, and estimated that 39.1462 animals were killed annually on each mile of road at that time.

One of the earliest works on the subject was *Feathers and Fur on the Turnpike*, a modest little book published in the late 1930s and which cited surprisingly high roadkill figures in New England. A private study of Nebraska animal road-casualties conducted in the early 1940s resulted in some 77,000 miles of travel and 6723 flattened animals, and showed that even before America really took off into the Automotive Age in the 1950s, the roadkill question was of some concern to at least a few biologists and researchers. But not, apparently, to the vast majority of Americans, even today, when the wildlife road carnage is very much higher.

All of this, plus the sheer increase in traffic speed and volume, as well as the growing overall roadway mileage, has combined to produce the estimated one million daily animal casualties, at least 600,000 of them birds.

Immediate solutions to the roadkill dilemma are hard to come by, involving, as they most surely will, large and costly adjustments to the status quo where it comes to personal mobility. Among the few that appear to be at least within the realm of workability, but easier said than done, are:

- The further development of effective animal detection aids, both on roadways and on vehicles.
- The blocking or deterring of animals from entering or attempting to cross roadways.
- The development of better headlight illumination to allow drivers to see animals far ahead in the roadway.
- Driver education in the skills required to watch for and avoid animals on roads.

All of these proposals, as well as ambitious but impractical "animal underpasses and overpasses," false predator scents sprinkled along roadsides, and sophisticated electronic fencings, offer the best of intentions in the campaign to stem the wildlife road slaughter. But it is doubtful that physical animal-crossing barriers, for example, could be installed along enough miles of roadways to produce a meaningful reduction of the carnage without an immense expenditure of public funds—and at what cost to the esthetics of the roadside? Will an intensified driver education program have much impact on a proliferating army of drivers subjected to relentless and persuasive advertising for more powerful, ego-boosting, and faster vehicles? The simple fact is that our roadways and the vehicles that ply them were not designed to function with unpredictable animal movements in mind. And neither, apparently, was the human driver.

The prospect for the resolution of the roadkill problem is far from bright, for as long as oil supplies remain plentiful and relatively cheap there seems little likelihood that the world will abandon its love affair with the internal combustion engine and the motor vehicles it powers. In view of this the roadkill toll seems destined to continue to rise, for at least as long as viable habitat for the production of victims exists between the traffic arteries of the vast transportation system that today spans the continent. Ultimately, however, the finite supply of fossil fuels will most assuredly run dry and blissfully mobile humankind just might find itself sitting at a horse's tail once more, or perhaps on the seat of a leg-powered bicycle.

The lowly bike has long been relegated to the status of plaything even though it is the principal means of human transport in most Third World nations, and more than 100 million of them are manufactured worldwide each year—about three times the number of new cars. The bicycle has long been recognized as a contrivance whose recreational use promotes overall health and well-being, and this indeed is the use and purpose to which some 100 million bikes in the United States are put. Although the U.S. is second only to China in bike ownership and has twice as many bicycles as India, very few bicycle owners in North America commute to work on them or pedal to the local supermarket. Nor is any real provision made in the road systems of our larger cities that would enable them to do so with any degree of safety. In China, conversely, there are more than 275 million bicycles, the great majority of which serve as practical and vital daily transport for their owners and their goods. China, however, is still home to something less than a million motor vehicles, of which relatively few are in private ownership.

A "kinder and gentler" world road system traveled solely by bicyclists is not likely to appear in the near future; the immensely powerful automobile and highway lobby will see to that as long as the supply of fossil fuels lasts. While more than 40 million acres have been paved over in the U.S. in the form of roads and parking lots to accommodate the growing auto horde, little or no thought is given to providing safe passage for bicycles. On most major highways their use is forbidden as too dangerous.

But changes are in the polluted wind of many cities, where auto congestion gridlock and rising commuting costs are encouraging more and more people to switch to cycling whenever they can practically do so. The number of bicycle commuters in the U.S. reached 1.9 million in the late 1980s. This is a mere two percent of all commuters, but the rise does represent a quadrupling in a little over a decade, and just might presage the day when a deer, a raccoon, or the family cat might actually make it across a busy road alive. It all depends on how quickly bicycle use graduates from being a personal choice to an urgent item on the public agenda; only then might the motor vehicle cease to be the tyrant it has become, and the ubiquitous roadways become less the malevolent death traps they now are for wildlife.

Animals enter and cross roadways for a number of reasons: mammals seek the salt residues left over from winter road maintenance; birds are attracted to sun-warmed roads in order to feed on insects, animal carcasses, and worms; snakes and other reptiles bask on the warm, post-dusk pavements; fast foods and other organic litter discarded along our roadways is irresistible to omnivorous creatures; and many wild animals attempt crossings in the course of their daily wanderings to establish new territories, escape predation, or to seek new food sources or mates. The animal death toll is traditionally higher in spring, when migrating birds pass through a region, reptiles and amphibians emerge from hibernation, and resident mammals respond to their amorous urges with attacks of wanderlust. Although numberless raccoons, armadillos, muskrats, deer, pheasants, a careless starling or two, and other assorted wildlife survivors meet their ends on the increasingly busy roadways of North America, for the most part these and all of the other adaptable species that qualify for wildlife survivorship have come to recognize both the assets and hazards of the modern roadway. Whether these "street smarts" will do our urbanized wildlife any good over the long run, as the great conurbations spread ever further into the countryside in the next century, remains to be seen.

While the roadkill problem might appear to be a social and environmental challenge of daunting and virtually insoluble proportions, that is not to say that there is a shortage of people willing to rise to the task, no matter how intimidating. A national anti-roadkills task force has been formed "to facilitate the creation and deployment of public education drives, compile reliable ethological data on which to base social policy, and to lobby automakers and transportation officials" in the search for a solution to the growing wildlife carnage on the roads. Readers interested in supporting or joining this effort can contact the National Anti-Roadkills Project, c/o VNN, Box 68, Westport, CT 06881, (203) 452-7655.

Habitat preservation

The universe inhales, a snowflake falls,
galaxies collide, a flower blooms, fades,
a bird sings, the cosmos reels,
autumn leaves turn
death and life
eternal duet.

After a year, a decade, a century has passed
we look back in wonder at the beauty

of time's tapestry.
We live, we love, we praise God, we die
the universe exhales.

Joseph R. Veneroso Maryknoll, January 1992

Eight of the nation's largest conservation groups have drafted a resolution calling for action to save the planet's oceans from further degradation and destruction. The resolution outlines the following six solutions to the growing human threat to the health of the world ocean:

1. Enactment of a federal oceans and coastal-protection policy to direct national legislation in a comprehensive and coordinated effort toward reducing coastal pollution and restoring estuaries, wetlands and fisheries.
2. Strong enforcement of and adequate funding for the goals outlined in the Clean Water Act, the Endangered Species Act, the Toxic Substances Control Act, the Coastal Zone Management Act, Federal Insecticide, Fungicide, and Rodenticide Act and other laws relating to coastal protection.
3. Funding for coastal and oceans research, including alternative systems for sewage and waste dumping, and incentives for reducing the flow of industrial toxins into coastal waters, as well as creation and enforcement of regulations for handling, transporting, and disposing of toxins.
4. Enactment of new, comprehensive legislation designating and protecting ecologically sensitive and important areas of the oceans for scientific study and for the enjoyment of future generations.
5. Support of efforts by coastal states to ban offshore development in environmentally and economically sensitive areas where it would threaten beaches, coastal communities, and commercial and recreational fishing.
6. Support of congressional deferral of outer-continental shelf oil development in unexploited areas pending a complete review of existing oil-development laws and regulations coupled with creation of a national energy policy that emphasizes clean, safe, renewable, and efficient energy sources.

Making a difference

Since the first human eye saw a leaf in Devonian sandstone and a puzzled finger reached to touch it, sadness has lain over the heart of man. By this tenuous thread of living protoplasm, stretching backward into time, we are linked forever to lost beaches whose sands have long since hardened into stone. The stars that caught our blind amphibian stare have shifted far or vanished in their courses, but still that naked, glistening thread winds onward. No one knows the secret of its beginning or its end. Its forms are phantoms. The thread alone is real; the thread is life.

Loren Eiseley, *The Firmament of Time* (Atheneum, 1960)

The key to "making a difference" when it comes to improving the lot of the world we live in and the creatures that share it with us lies in personal involvement, no matter how modest the level. Thomas Paine once remarked that "evil prevails because good men do nothing." The ongoing destruction of the earth's biosphere does not, in my opinion, constitute an evil, but rather simply a consequence of humankind's collective failure to live up to the label of sapiens, or "thinker" it so readily and gratuitously assigned itself.

There is not, indeed, a great deal the average individual can do to stem the tide of environmental decline taking place on earth today. That task belongs to the relative handful of humans who hold the international reins of power and dictate policy to the rest of us. But each small positive act or willingness to sacrifice a little comfort and convenience on behalf of the planet and its creatures does make a difference when multiplied many times. "A thousand points of light," someone once said.

The simple act of taking a plastic trash bag along on a trip to the beach or woods and stuffing it with all of the litter you see won't make any real dent in the billions of pounds of the stuff desecrating the world landscape, but it assuredly will make you feel a better person for having done it. And sometimes, in the face of monumental odds, that small sense of personal commitment can be the most important victory of all. The following section offers a few tips on pitching in on the personal and local level.

Community action

- Question every new development or building program in your locality as to its environmental consequences and real economic and social impact. We should be more concerned for the living landscape than for the pocketbooks of the developers.
- Landowners cannot be allowed to do anything they wish with their land. Consideration of the surrounding populace and landscape should be foremost.
- Create and enforce a general plan for your community. Without an idea of what direction development must take to satisfy the community's economic and environmental needs, disorganized and detrimental outward expansion will soon destroy the potential of the region to achieve a healthful accord of man and nature.
- Support the strength of planning laws. A general plan can be effective only if it is strongly enforced. Zoning land is a relatively weak way to protect it. Zoning changes are too commonly approved by the very people who planned them to protect certain areas.
- Reorganize taxes so that agricultural land will not be so susceptible to development. If our farmlands were not taxed out of existence, there would be less loss of these valuable, food-producing open spaces.
- Federal activities, such as building dams, plants, and military bases can have a great impact on an area and might attract thousands of new occupants. Before encouraging and allowing Federal construction in your area, consider the impact of these extra people.

Epilogue

Will our grandchildren ever know the sudden glory of a dawn in an unpolluted sky or witness the slow fading of the violet light that floods a clear twilight heaven? Will they explore mountain meadow knee deep in columbine or walk the tranquil aisles of a virgin forest where the trees soar tall and proud to meet the Sun? As these sights and sounds are replaced by cold stone and hard steel and plumes of acrid smoke, then these will become part of the child. A fabric woven of such coarse threads will make a harsher man.

> **Louise B. Young**
> *Sowing the Wind* (Prentice-Hall, 1990)

It had been a very long day. Although writing about the natural world and the creatures that inhabit it is not an unpleasant activity, it can be a demanding one, and the unwinking gaze of a little laptop computer will eventually play havoc with the eyes. Emerging from the tunnel of concentration, I saw that evening had come. I leaned back in the old chair, rubbed my forehead and found myself listening to the rich, "day-is-done" caroling of a robin somewhere out in the surrounding dusk. The woods just outside our little house in the mountains of Virginia were silent, cool, and deep purple-green, awaiting the night.

On sudden impulse, I went out into the gathering dark to take stock of the world as it existed here and now in these ancient, rumpled hills, weathered monadnocks that were middle-aged when the dinosaurs walked the alien continents of the Mesozoic.

I stood in the fading light, in the lush, unruly sward at the edge of our large garden at the edge of the forest and took in the scene around me. To the west, the burnished bronze sky spoke of the descent of the dying sun, now on its way to the other side of the world. Eastward, the massive, dark rampart of Seven Mile Mountain rose against the graying indigo of the new night sky; a full moon hung in the void above the ridge, cool, marbly white against the sky. A bat flickered past, high, high up, a darting, ethereal mote against the blank page of the descending night, and then was gone; good . . . life yet moved about in the heavens. The silence of the place was complete.

Ah, but not quite. I became aware of the little sounds of the hidden world that exists at our feet. Crickets strummed and chirred in the worn and dusty grasses of the late-summer garden; something rustled in the weeds at the edge of the dry and spent soil. But that was all.

I stood and wondered what the nights here must have looked and sounded and smelled like thirty, fifty, maybe a hundred years ago. Surely this time of evening would have resounded with a cacophony of owls and night frogs and the stirrings of a hundred times a hundred unnamed little things moving about on their desperate errands of survival. Surely the sky would have been alive with great looping and flapping night moths, pale against the gloom, and the fireflies must have been as summer blizzards against the darkening wall of the forest. I'd never know, for I was a newcomer here in these ancient and timeless hills. My neighbors often spoke of the past abundance of this or that bird or mammal, but even they, who had lived here all their lives, were hard-pressed to define the difference between then and now in any meaningful way. As for me, I would have to accept the experience of standing alone, perhaps literally, in silence on a late-summer night under the ageless stars, as it came to me now, in the nineteen hundred and ninety-second year of this, the Second Millennium.

As I stood beneath the night sky, a kind of "watcher at the fen," the cricket chorus ratcheted into lower gear and finally slipped into silence as one by one its hidden members fell mute. A small, almost casual, leaflike flicker of movement caught my eye, near the woods at the far end of the garden. I peered through my glasses with civilized, night-blind eyes and tried to separate fact from fantasy in the gloomy blurrings and dim forms that composed the ancient nocturne coming to life before me. I stared and stared with frustrated intensity, and at last the pieces of the puzzle slipped and slid together and became the washy-brownish, indistinct form of a deer. It had been there all along, gazing impassively yet alertly at me, its working mouth filled with produce clippings, as I sought to penetrate the darkness and see the things that were right there, in front of me. The deer flicked an ear again, then lowered its head and smoothly tugged and clipped the lank tomato vines. Another materialized from the confusion of woods-edge vegetation, soundlessly stepped across the rows, cropped and ate, head up, head down, ear flicks, dark eyes watching, wary yet bold.

Relaxing, I shifted ever so slightly, a twig popped underfoot. The brownish shadows moved, shifted just a hair's breadth to the side and were gone; they had simply melted back into the scrim of trees as silently as they had arrived. Here, in the mountains, men meant guns, and the deer were taking no chances. I was carrying nothing but a small, household flashlight, but I chose not to use it against the beasts.

Exhaling slowly, I shivered, for a light wind had moved down the valley and tousled the drying ranks of corn and snap beans. The air grew cold and oddly wintry, though fall was still weeks away. The planet moved beneath me, its grassy hide a labyrinth of earthen pathways peopled by vast, hidden armies of insect gremlins and countless billions of microbes, all busy converting the stuff of life into the stuff of the globe itself. Scuffing the drying turf, I couldn't see or hear them, of course, for they moved in minute ways beyond my perception. To my senses, the valley appeared empty and void of motion, and thus of animated life. I mourned this, as the wind swept the silent sky and the old moon rose whitely and neared the zenith. Surely, the wildlings would be abroad on such a lush, dark-green night filled with whispers as this.

But then I heard it again, that swift, furtive rustling just beyond the edge of the garden. Something was moving with a fierce, rushing intensity, pursuing something else; there in the dim, grassy tangles, a life-and-death pursuit was taking place. I strained to see, and a small, smokey-gray form took shape, a pale, transitory night-thing slipping deliberately, yet with great stealth, through the miniature jungle of Queen Anne's Lace, mullein, and briars. A wraith of fog? An illusion of weary eyes?

With a furious rush and rustle, the pounce came, followed by the high, wire-thin death cry of a small creature. Further leafy scuffling, and then a cat made a wonderful, arcing leap out of the grass and landed with controlled and fluid grace on the cropped sod of the lawn, not twenty feet from my wondering eyes.

I pointed the flashlight and snapped it on. It was one of my neighbor's animals; a coarse-coated, black and white tom, and it had a small animal in its mouth. I walked closer and the cat froze and stood its ground. The luckless mouse that was indeed the victim flipped about and shivered its tail spasmodically. The cat, sensing competition for its hard-won prey, crouched low and growled softly around the body of the mouse. Curious as to just what rodent species had fallen to the cat's hunting prowess, I approached still closer and the cat backed slowly away, into the cover of the weeds, its eyes blazing from its concealment with reflected ferocity in the light's feeble beam. I opted for backing off, for I was close enough to note that the mouse was gray, with a grayish white underside; it was a house mouse, probably one of the several that lived in the heap of firewood in our small storage shed.

And as we paused there, frozen in tableaux within a fleeting moment of time in a dark valley under a silent, starred sky, I came to the distinctly uneasy realization that the three of us— the domestic cat, the house mouse, and the human being—were among the relatively few creatures that would, with reasonable certainty, still be prowling the nocturnal pathways of this mountain valley thirty, fifty, or maybe a hundred years hence. Here, on this night, three of the mammalian inheritors of the "new earth" had come together for one brief moment, while elsewhere over the valley, across the continent and the very planet itself, multitudes of other, less persistent organisms and species were numbering the days they had left to them. In the end, as we humans look on with resigned and uncomprehending bewilderment as the "old earth" is eroded away by our own numbers, it will all come down to this. To the victor belongs the spoils. To the cat and the mouse and the rat and the roach will belong the world of tomorrow.

I left the hunter to its prey, turned and walked slowly back to my little white house, standing there in a cleared patch of forest, bathed in cold moonlight and filled with the creature comforts I require for my own survival. A cricket chirped tentatively from beneath the wooden steps; it ceased as I placed my foot upon the first tread. I pondered the nature of the world awaiting my grandchildren and their children and theirs, one that I would not live to see, but had gained some small insight into tonight. I went inside, into the artificial light and warmth of my house, and closed the door against the weedy rustlings, the whispered rodent voices, and the silent sky outside.

Environmental organizations

The Blue Water Task Force
Surfrider Foundation
(800) 743-SURF

Global Action Plan for the Earth
84 Yerry Hill Road
Woodstock, NY 12498
(914) 679-4830

Alliance to Save Energy
1725 K Street, NW
Washington, DC 20006
(202) 857-0666

American Forestry Association
P.O. Box 2000
Washington, DC 20013
(202) 667-3300

Aquatic Conservation Network
540 Roosevelt Avenue
Ottawa, Ontario
CANADA K2A 128

Environmental Defense Fund
257 Park Avenue South
New York, NY 10010
(212) 505-2100

Friends of the Earth
530 7th Street, SE
Washington, DC 20003
(202) 544-2600

Greenhouse Crisis Foundation
1130 17 Street, NW
Washington, DC 20036
(202) 466-2823

Greenpeace USA
1436 U Street, NW
Washington, DC 20009
(202) 462-1177

League of Conservation Voters
320 4th Street, NE
Washington, DC 20002
(202) 785-8683

National Wildlife Federation
1400 16th Street NW
Washington, DC 20036
(202) 797-6800

Natural Resources Defense Council
40 West 20th Street
New York, NY 10011
(212) 727-2700

7The Nature Conservancy
1815 North Lynn Street
Arlington, VA 22209
(703) 841-5300

Rainforest Action Network
300 Broadway, Suite 28
San Francisco, CA 94133
(415) 398-2732

Rainforest Alliance
270 Lafayette Street, Suite 512
New York, NY 10012
(212) 941-1900

Sierra Club
730 Polk Street
San Francisco, CA 94109
(415) 776-2211

The Wilderness Society
900 17th Street SW
Washington, DC 20006
(202) 833-2300

Wilderness Awareness School
(education)
P.O. Box 755
Lincroft, NJ 07738
(908) 530-7717

Worldwatch Institute
1776 Massachusetts Avenue NW
Washington, DC 20036
(202) 452-1999

World Wildlife Fund
1250 24th Street, NW
Washington, DC 20037
(202) 293-4800

Bibliography

Allen, Mea, *Weeds*, New York: The Viking Press, 1978.

Anderson, Sherwood, *Hometown: The Face of America*, New York Alliance Books, 1940.

Arnett, Ross H. & Jacques, Richard L., *Simon & Schuster's Guide to Insects*, New York: Simon & Schuster, 1981.

Asimov, Isaac, and Pohl, Frederik, *Our Angry Earth*, New York: Tom Doherty Associates, 1991.

Ballantine, Bill, *Nobody Loves a Cockroach*, Boston: Little, Brown & Co., 1968.

Bare, Colleen Stanley, *Tree Squirrels*, New York: Dodd & Mead, 1983.

Bellini, James, *High Tech Holocaust*, San Francisco: Sierra Club Books, 1989.

Borror and White, *Insects (American North of Mexico)*, Boston: Houghton Mifflin Co., 1968.

Brown, Lauren, *Weeds in Winter*, New York: W.W. Norton, 1976.

Brown, Lester R., *The World Watch Reader*, New York: W.W. Norton, 1991.

Burt, William Henry, *A Field Guide to the Mammals*, Boston: Houghton Mifflin Co., 1952.

Burton, Maurice and Robert, *The World's Disappearing Wildlife*, New York: Marshall Cavendish Corp., 1978.

Caduto, Michael, and Bruchac, Joseph, *Keepers of the Animals*, Golden, Colo.: Fulcrum Publishing, 1991.

Caillet, Greg, et al. *Everyman's Guide to Ecological Living*, New York: Macmillan, 1971.

Chisholm, Anne, *Philosophers of the Earth*, New York: E.P. Dutton, 1972.

Curtis, Jane and Will, *Welcome the Birds to Your Home*, Brattleboro, Vt.: The Stephen Greene Press, 1980.

Davison, Verne, *Attracting Birds: from the Prairies to the Atlantic*, New York: Thomas Y. Crowell, 1967.

Dubos, Rene, *So Human an Animal*, New York: Scribners, 1968.

Ehrlich, Paul R. & Anne H., *Healing the Planet*, New York: Addison-Wesley, 1991.

Eiseley, Loren, *The Firmament of Time*, New York: Atheneum, 1960.

Erickson, Jon, *Dying Planet*, Blue Ridge Summit, Pa.: TAB Books, 1991.

Fichter, George S., *Insect Pests*, New York: Golden Press, 1966.

Friedenberg, Daniel, *Life, Liberty, and the Pursuit of Land*, Buffalo: Prometheus Books, 1992.

Gill, Don, and Bonnett, Penelope, *Naturel in the Urban Landscape: A Study of City Ecosystems*, Baltimore: York Press, 1973.

Grimm, William Carey, *The Book of Trees*, Harrisburg, Pa.: Stackpole Co., 1962.

Guthrie, D.M. & Tindall, A.R., *The Biology of the Cockroach*, London: Edward Arnold, Ltd, 1968.

Harris, Charles, *Eat the Weeds*, Barre, Mass.: Barre Publishers, 1961.

Headstrom, Richard, *Suburban Wildlife*, Englewood Cliffs, N.J.: Prentice-Hall, 1984.

Helmer, John and Eddington, Neil A. eds., *Urbanman*, New York: The Free Press, 1973.

Hoage, R.J. ed., *Animal Extinctions*, Washington, D.C.: Smithsonian Institution Press, 1985.

Hocking, Brian, *Six-legged Science*, Cambridge, Mass.: Schenkman Publishing Co. Inc., 1968.

Kelley, Ben, *The Pavers and the Paved*, New York: Donald W. Brown, 1971.

Kinkead, Eugene, *Wildness Is All Around Us*, New York: E.P. Dutton, 1978.

Knutson, Roger M., *Flattened Fauna*, Berkeley, Calif.: Ten Speed Press, 1987.

Lamb, Robert, *World Without Trees*, New York: Paddington Press, 1979.

Lawrence, Gale, *The Indoor Naturalist*, New York: Prentice-Hall, 1986.

Laycock, George, *The Alien Animals*, New York: The Natural History Press, 1966.

Luoma, Jon R., *Troubled Skies, Troubled Waters*, New York: The Viking Press, 1984.

Lutz, Frank E., *Fieldbook of Insects*, New York: G. P. Putnam's Sons, 1918.

MacClintock, Dorcas, *Squirrels of North America*, New York: Van Nostrand Reinhold, 1970.

Manes, Christopher, *Green Rage*, Boston: Little, Brown, 1990.

Mathews, Dick, *Wild Animals as Pets*, Garden City, N.Y.: Doubleday & Co, Inc., 1971.

Mathews, F. Schuyler, *Field Book of American Trees and Shrubs*, New York: G.P. Putnam's Sons, 1915, 25th ed.

Matthiessen, Peter, *Wildlife in America*, New York: The Viking Press, 1959.

McLoughlin, John C., *The Canine Clan; A New Look at Man's Best Friend*, New York: Viking Press, 1983.

McMillon, Bill, *Nature Nearby*, New York: John Wiley & Sons, 1990.

Mills, Stephanie, ed., *In Praise of Nature*, Washington, D.C.: Island Press, 1990.

Mitchell, George J., *World on Fire*, New York: Scribners, 1991.

Mohlenbrock, Robert H., *Where Have All The Wildflowers Gone?*, New York: MacMillan, 1983.

Nash, Hugh, ed., *Progress as if Survival Mattered*, San Francisco: Friends of the Earth, 1977.

Orians, Gordon, *Blackbirds of the Americas*, Seattle: University of Washington Press, 1985.

Peterson, Roger Tory, *A Field Guide to the Birds*, Boston: Houghton Mifflin Company, 1934, revised.

Phillips. Roger, *Trees of North America and Europe*, New York: Random House, 1978.

Rae, John B., *The Road and Car in American Life*, Cambridge, Mass.: MIT Press, 1971.

Rathje, William, & Murphy, Cullen, *Rubbish! The Archaeology of Garbage*, New York: Harper Collins, 1992.

Revkin, Andrew, *Global Warming: Understanding the Forecast*, New York: Abbeville Press, 1992.

Ricciuti, Edward R., *Killer Animals*, New York: Walker & Co. 1976.

Ritchie, Carson I. A., *Insects, the Creeping Conquerers*, New York: Elsevier-Nelson Books, 1979.

Rood, Ronald, *Animals Nobody Loves*, Brattleboro, Vt.: The Stephen Greene Press, 1971.

Schneider, Stephen H., *Global Warming*, New York: Randon House, 1989.

Simmons, James R., *Feathers and Fur on the Turnpike*, Boston: Christopher Publishing House, 1938.

Simon, Anne W., *The Thin Edge: Coast and Man in Crisis*, New York: Harper & Row, 1978.

Snedigar, Robert, *Our Small Native Animals: Their Habits and Care*, New York: Dover Publications, 1963 reprint.

Spirn, Anne Whiston, *The Granite Garden: Urban Nature and Human Design*, New York: Basic Books, 1984.

Stebbins, Robert C., *A Field Guide to Western Reptiles and Amphibians*, Boston: Houghton Mifflin Co., 1985.

Stokes, Donald W., *A Guide to Observing Insect Lives*, Boston: Little, Brown, 1983.

Sunset Pub., Editors, *An Illustrated Guide to Attracting Birds*, Menlo Park, Calif.: Sunset Publishing Corp., 1990.

Teale, Edwin Way, *The Strange Lives of Familiar Insects*, New York: Dodd, Mead & Co., 1962.

Tudge, Colin, *Global Ecology*, New York: Oxford University Press, 1991.

Watkins, Jon William, *Suburban Wilderness*, New York: G. P. Putnam's Sons, 1981.

Watson, J. Wreford, *North America: Its Countries and Regions*, New York: Frederick A. Praeger, 1967.

Wenner, Jan, Ed., *The Rolling Stone Environmental Reader*, Washington, D.C.: Island Press, 1992.

Young, Louise B., *Sowing the Wind*, New York: Prentice-Hall, 1990.

Index

204

206